NOTES ON

MECHANISM ANALYSIS

PRESS, INC.

Prospect Heights, Illinois

For information about this book, write or call:

Waveland Press, Inc.
P.O. Box 400
Prospect Heights, Illinois 60070
(708) 634-0081

Copyright © 1981 by Balt Publishers
First published by Waveland Press, Inc., 1986

ISBN 0-88133-270-4

Printed in the United States of America

7 6 5 4

PREFACE

These notes were originally prepared as a supplement to excellent existing texts which emphasize graphical or semi-graphical methods for the analysis of mechanisms. The intent here is to point out methods which might lend themselves to computerized analyses.

The author's debt to others is difficult to document, but he would like especially to acknowledge the influence of Dr. J. Modrey, Dr. W. Soedel, Dr. R. Cipra, Dr. D. Tesar, Dr. C. Radcliffe, and Dr. B. Paul. Each of these people, through their publications and/or personal conversations, have affected the author's thinking about what should be taught and how to go about it. Others may also recognize some of their thoughts reflected here. For whatever is good I thank them. For whatever is not so good I take responsibility.

While these are considered supplementary notes for a full-scale course in mechanism analysis, they might stand alone for a shorter course. For a very short course one might use only chapters 1, 3 and 5.

A number of unsolved problems are included. The thought here is that these might be used as "class problems," i.e., the instructor might use these as the basis for some classroom discussion while the students take appropriate notes, hence the blank areas provided with these pages. Alternatively, some of these might be used as "home problems." Instructors are hereby granted permission to reproduce the problem pages for this purpose.

Allen S. Hall, Jr.
June 1981

CONTENTS

Chapter 2   KINEMATIC ANALYSIS II

Chapter 3   FORCE ANALYSIS I

CHAPTER 1

KINEMATIC ANALYSIS I

1.1 Introduction

Kinematic analysis solves the following problem. Given the state of motion
(position, velocity, acceleration) of one link (the "input" link). Determine
the state of motion (position, velocity, acceleration) of all other links and
particular points of interest. There are several ways of dealing with this
problem. Any basic mechanics course treats the problem as a preliminary to
problems in dynamics. Our approach is tailored to the use of modern computational aids. We call it the vector loop approach.

1.2 The Vector Loop Approach

Briefly, the vector loop approach consists of the following steps.

(1) Attach to the links of the mechanism vectors forming a closed loop
     (or loops).

(2) The vectors should be chosen and defined such that the variables
     (vector lengths or angles) are variables of interest in the
     mechanism. For example, if a chosen vector has a fixed length
     and a variable angle, the variable angle should measure the
     angular motion of one of the links.

(3) Write the vector-loop equation. This is simply the statement that the sum of the vectors in the loop is zero.

(4) Choose an X,Y coordinate system. Break the vector loop position equation into two scalar component equations. Solve these for the _position_ unknowns.

(5) Differentiate the position equations with respect to time. Solve these for the _velocity_ unknowns.

(6) Differentiate again with respect to time. Solve for the _acceleration_ unknowns.

There are a number of details involved in carrying out the above outline which we will explain and explore in the following series of examples.

1.3 4-Bar Linkage (Fig. 1.1)

Figure 1.1 illustrates the first step of the approach as applied to a simple mechanism, the 4-bar linkage. The indicated vectors are defined as follows:

$\bar{r}_2$, attached to link 2, from O to A.

$\bar{r}_3$, attached to link 3, from A to B.

$\bar{r}_4$, attached to link 4, from M to B.

$\bar{r}_1$, attached to link 1, from O to M.

The vector loop position equation is:

$$\bar{r}_2 + \bar{r}_3 - \bar{r}_4 - \bar{r}_1 = 0$$

This expresses the fact that this loop of vectors is closed, and will remain closed as motion takes place. The _constants_ in the equation are r_2, r_3, r_4, r_1 (link lengths) and θ_1 (angle to vector $\bar{r}_1$ from whatever reference is chosen). The _variables_ are θ_2, θ_3, θ_4 (angles to vectors $\bar{r}_2$, $\bar{r}_3$, $\bar{r}_4$ from the chosen reference).

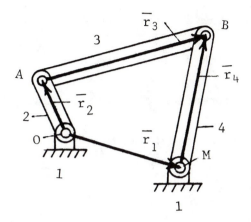

Fig. 1.1 Vector loop for 4-bar linkage.

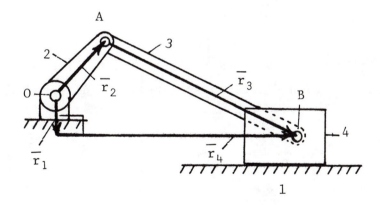

Fig. 1.2 Vector loop for off-set slider-crank.

1.3

1.4 Off-set Slider-Crank (Fig. 1.2)

The vectors are defined as follows:

$\bar{r}_2$, attached to link 2, from O to A.

$\bar{r}_3$, attached to link 3, from A to B.

$\bar{r}_4$, parallel to direction of motion of link 4.

$\bar{r}_1$, perpendicular to $\bar{r}_4$, from O.

The vector loop equation is:

$$\bar{r}_2 + \bar{r}_3 - \bar{r}_4 - \bar{r}_1 = 0$$

The constants are r_1, r_2, r_3, θ_4, θ_1. Variables are θ_2, θ_3, r_4.

1.5 A Two-Loop Mechanism (Fig. 1.3)

In many mechanisms a single vector loop is not adequate to describe the kinematic behavior. In Fig. 1.3, one loop is formed of vectors attached to the 4-bar linkage, links 1, 2, 3, 4.

$$\text{Loop I,} \quad \bar{r}_2 + \bar{r}_3 - \bar{r}_4 - \bar{r}_1 = 0 \tag{1}$$

A second loop is formed by vectors $\bar{r}_6$ (attached to link 6), $\bar{r}_7$ (attached to link 3 and perpendicular to the slot center line), $\bar{r}_5$ (directed along the slot center line), plus vectors $\bar{r}_4$ and $\bar{r}_8$ to close the loop.

$$\text{Loop II,} \quad \bar{r}_6 - \bar{r}_5 - \bar{r}_7 - \bar{r}_4 + \bar{r}_8 = 0 \tag{2}$$

The variables in loop I are θ_2, θ_3, θ_4. In loop II they are θ_6, θ_4, r_5 and θ_3. (Note that $\theta_7 = \theta_3 + \lambda$ and $\theta_5 = \theta_3 + \lambda - 90°$.)

Whether we have _enough loops_ and _good vectors_ can be determined by answering these two questions.

(1) Is the number of variable unknowns equal to the number of independent scalar equations?

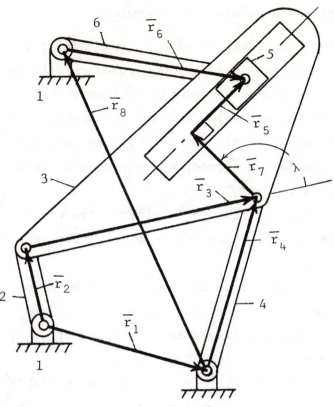

Fig. 1.3 A two-loop mechanism.

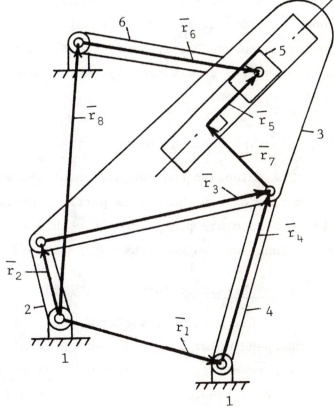

Fig. 1.4 An alternative loop
        formation.

1.5

(2) Do the variables include all those we wish to determine?

In this example we have two vector loop equations, which mean _four_ scalar position equations. The total number of variables is five. One of these will be
the "input" variable (known), hence there are _four_ variable unknowns. The
answer to the first question is "yes." The variables measure the angular
motions of all the links and also the motion of block 5 relative to link 3.
These are all the mechanism motion variables we should have any need to determine. The answer to the second question is "yes."

In any multi-loop mechanism there will be acceptable alternative ways of
forming the needed loops. For example, see Fig. 1.4 for an alternative way of
forming a second loop for this same mechanism.

It should be observed that, for a multiloop mechanism, it will sometimes
be possible to solve the loops separately but sometimes they must be solved
simultaneously. In the example of this article, if θ_2 is the input variable,
then loop I contains only two variable unknowns, θ_3 and θ_4, hence can be solved
by itself. After loop I is solved loop II will contain only the two variable
unknowns, θ_6 and r_5, so can be solved. However, if θ_6 were the input variable,
loop I would contain three variable unknowns, θ_2, θ_3 and θ_4, and loop II would
contain unknowns θ_4, r_4, and θ_2. Neither loop could be solved alone. The two
must be solved simultaneously, which means the simultaneous solution of four
scalar equations.

1.6 Choosing Vectors, Hints

The selection and definition of the most appropriate vectors to describe
the motion of a mechanism is partly a matter of practice. If all the joints of
the mechanism are pin joints, the choices will be fairly obvious, but for other
situations some suggestions may be in order.

Straight-slider joint (Fig. 1.5)

Use a pair of vectors, one in the direction of sliding and one
perpendicular to the direction of sliding. For example, vectors $\bar{r}_8$
and $\bar{r}_7$ of Fig. 1.5. Vector $\bar{r}_7$ is of fixed length, but turns with
link 7. Vector $\bar{r}_8$ varies in length (measuring the relative motion
between links 8 and 7) and also turns with link 7.

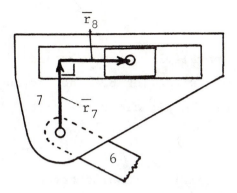

Fig. 1.5 Vectors associated with slider joint.

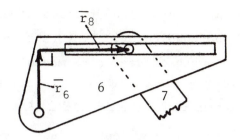

Fig. 1.6 Vectors associated with pin-in-straight-slot joint.

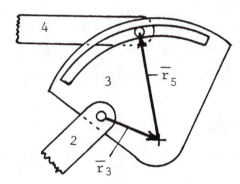

Fig. 1.7 Vectors associated with pin-in-circular-slot joint.

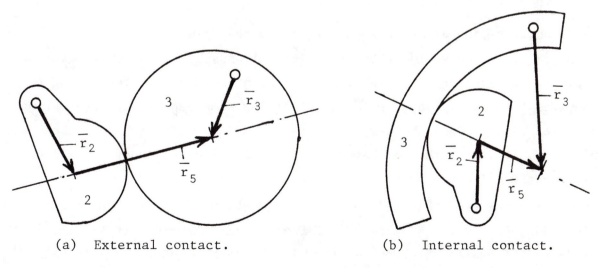

(a) External contact.

(b) Internal contact.

Fig. 1.8 Vectors associated with contact of circular bodies.

1.7

<u>Pin in a straight slot</u> (Fig. 1.6)

Much like Fig. 1.5. Use one vector parallel to the slot and one perpendicular. Vectors $\bar{r}_6$ and $\bar{r}_8$ both turn with link 6. The varying length of vector $\bar{r}_8$ measures the motion of the pin along the slot.

<u>Pin in a circular slot</u> (Fig. 1.7)

In the loop include a vector ($\bar{r}_5$ in Fig. 1.7) running between the pin and the center of curvature of the slot. This is a vector of fixed length, but varying angle.

<u>Circular bodies in contact</u> (Fig. 1.8)

Include a vector running between the centers of curvature of the two bodies, for example, vector $\bar{r}_5$ in Fig. 1.8(a) or (b). This is a vector of fixed length, but varying angle.

1.7   Coordinate System Choice

In our outline of the vector loop approach to kinematic analysis of mechanisms we listed step (4) as

(4)   Choose a fixed X,Y coordinate system. Break the vector loop position equation into two scalar component equations. Solve these for the <u>position</u> unknowns.

The choice of coordinate system origin and the orientation of the axes relative to the mechanism is arbitrary. It will usually be convenient to use a fixed pivot location for the origin, and to align the X axis either along the line between two fixed pivots or in the direction some especially interesting motion takes place. Good choices for the 4-bar and slider-crank mechanisms are shown in Fig. 1.9. For the mechanism of Fig. 1.10 we might choose to align the X axis with the direction of travel of the output slider.

We will use a right-hand coordinate system and measure all angles positive counterclockwise. In particular, the <u>angular positions</u> of all vectors will be described by the angles "θ" measured <u>from</u> the positive X axis <u>counterclockwise</u> to the direction of the vector.

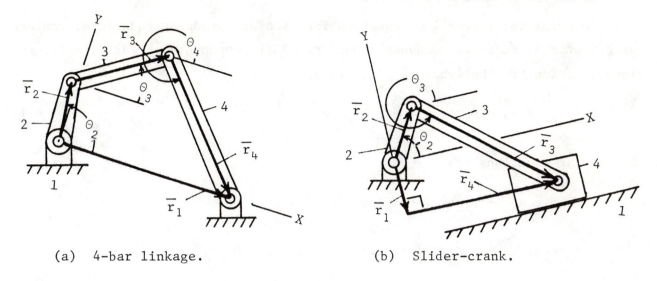

(a) 4-bar linkage.

(b) Slider-crank.

Fig. 1.9   Coordinate systems.

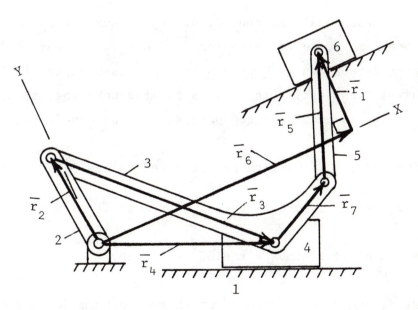

Fig. 1.10   Vector loops and coordinate system for a six-link mechanism.

1.8 Position Solution, Iterative

To break the vector loop equation into scalar component equations, replace each vector by $r\cos\theta$ (X components) and $r\sin\theta$ (Y components). For example, the loop equation for the 4-bar of Fig. 1.9(a) is

$$\bar{r}_2 + \bar{r}_3 + \bar{r}_4 - \bar{r}_1 = 0 \tag{1}$$

The scalar equations are

$$r_2\cos\theta_2 + r_3\cos\theta_3 + r_4\cos\theta_4 - r_1\cos\theta_1 = 0 \tag{2}$$

$$r_2\sin\theta_2 + r_3\sin\theta_3 + r_4\sin\theta_4 - r_1\sin\theta_1 = 0 \tag{3}$$

We chose the X axis to be directed along the vector $\bar{r}_1$, angle $\theta_1 = 0°$. We may as well recognize this immediately and simplify our equations accordingly.

$$r_2\cos\theta_2 + r_3\cos\theta_3 + r_4\cos\theta_4 - r_1 = 0 \tag{4}$$

$$r_2\sin\theta_2 + r_3\sin\theta_3 + r_4\sin\theta_4 = 0 \tag{5}$$

The variables in these equations are θ_2, θ_3, θ_4. Assuming θ_2 to be the "input" variable, the problem now is to solve for θ_3 and θ_4. One way to do this is by an iterative procedure known as Newton's method, based on the following reasoning which involves first estimating values for the unknowns, then systematically correcting those values until desired precision is obtained. We reason as follows.

(1) Estimate θ_3, θ_4. Call these estimated values θ_3', θ_4'. Then,

$$\varepsilon_1 = r_2\cos\theta_2 + r_3\cos\theta_3' + r_4\cos\theta_4' - r_1\cos\theta_1 \tag{6}$$

$$\varepsilon_2 = r_2\sin\theta_2 + r_3\sin\theta_3' + r_4\sin\theta_4' - r_1\sin\theta_1 \tag{7}$$

where ε_1 and ε_2 are "errors" which are to be made to approach zero by adjusting the values of θ_3' and θ_4'.

(2) To calculate how much adjustment to make we use the following approximations.

$$\Delta\varepsilon_1 \approx \frac{\partial\varepsilon_1}{\partial\theta_3'}\Delta\theta_3' + \frac{\partial\varepsilon_1}{\partial\theta_4'}\Delta\theta_4' \tag{8}$$

$$\Delta\varepsilon_2 \approx \frac{\partial\varepsilon_2}{\partial\theta_3'}\Delta\theta_3' + \frac{\partial\varepsilon_2}{\partial\theta_4'}\Delta\theta_4' \tag{9}$$

$$\frac{\partial\varepsilon_1}{\partial\theta_3'} = - r_3\sin\theta_3' \tag{10}$$

$$\frac{\partial\varepsilon_1}{\partial\theta_4'} = - r_4\sin\theta_4' \tag{11}$$

$$\frac{\partial\varepsilon_2}{\partial\theta_3'} = r_3\cos\theta_3' \tag{12}$$

$$\frac{\partial\varepsilon_2}{\partial\theta_4'} = r_4\cos\theta_4' \tag{13}$$

Hence,

$$\Delta\varepsilon_1 \approx (-r_3\sin\theta_3')\Delta\theta_3' + (-r_4\sin\theta_4')\Delta\theta_4' \tag{14}$$

$$\Delta\varepsilon_2 \approx (r_3\cos\theta_3')\Delta\theta_3' + (r_4\cos\theta_4')\Delta\theta_4' \tag{15}$$

(3) We will use equations (14) and (15) to calculate "corrections" $\Delta\theta_3'$ and $\Delta\theta_4'$. The calculating routine will be as follows.

(a) Estimate θ_3' and θ_4'.

(b) Compute ε_1 and ε_2 (equations (6) and (7)).

(c) Set $\Delta\varepsilon_1 = -\varepsilon_1$ and $\Delta\varepsilon_2 = -\varepsilon_2$.

(d) Solve for $\Delta\theta_3'$ and $\Delta\theta_4'$ (equations (14) and (15)).

(e)   Add $\Delta\theta_3'$ and $\Delta\theta_4'$ to the previous values of θ_3' and θ_4'
and return to step (b). Repeat until $\Delta\theta_3'$ and $\Delta\theta_4'$
become as small as desired. The then existing values
of θ_3' and θ_4' will be the desired solutions of scalar
equations (4) and (5) (or of vector equation (1)).

The process described above calls for estimates of θ_3 and θ_4 at each value
of input, θ_2, for which the solution is to be made. It is suggested that the
estimates in the first position of the mechanism be obtained from a scale drawing of the mechanism in that position. After the solution is completed for that
position, and the input is stepped to a new value $\theta_{2_{new}} = \theta_{2_{old}} + \Delta\theta_2$, the new
estimates of θ_3 and θ_4 can be the values just computed for the previous position. This works well if the "step," $\Delta\theta_2$, is not too large. Values of
approximately 10 degrees for $\Delta\theta_2$ will ordinarily be quite satisfactory.

The computation process will normally converge to an accuracy of 0.01 deg.
in 3 or 4 iterations.

1.9   Sample Problem, Position Solution by Iteration, Slider-Crank Mechanism

Refer to Fig. 1.11.

Position equations

$$r_2\cos\theta_2 + r_3\cos\theta_3 - r_4 = 0 \tag{1}$$

$$r_2\sin\theta_2 + r_3\sin\theta_3 + r_1 = 0 \tag{2}$$

Reasoning

Let θ_3' be an estimate of θ_3

r_4' be an estimate of r_4

Then

$$r_2\cos\theta_2 + r_3\cos\theta_3' - r_4' = \epsilon_1 \tag{3}$$

$$r_2\sin\theta_2 + r_3\sin\theta_3' + r_1 = \epsilon_2 \tag{4}$$

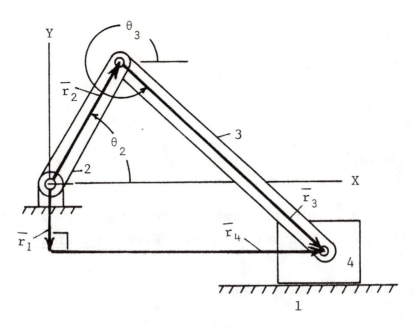

Fig. 1.11   Slider-crank for example problem. $r_1 = 1$ in., $r_2 = 2$ in., $r_3 = 4$ in., $\theta_2 = 60$ deg.

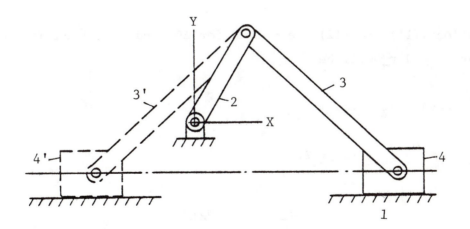

Fig. 1.12   Meaning of the two solutions to the position equations.

$$\partial \varepsilon_1 / \partial \theta_3' = -r_3 \sin \theta_3' \tag{5}$$

$$\partial \varepsilon_1 / \partial r_4' = -1 \tag{6}$$

$$\partial \varepsilon_2 / \partial \theta_3' = r_3 \cos \theta_3' \tag{7}$$

$$\partial \varepsilon_2 / \partial r_4' = 0 \tag{8}$$

If the estimated values of θ_3 and r_4 were changed by $\Delta\theta_3$ and Δr_4, respectively, then the "errors," ε_1 and ε_2, would change approximately as follows.

$$\Delta\varepsilon_1 = (-r_3 \sin\theta_3')\Delta\theta_3 + (-1)\Delta r_4 \tag{9}$$

$$\Delta\varepsilon_2 = (r_3 \cos\theta_3')\Delta\theta_3 + (0)\Delta r_4 \tag{10}$$

To make the errors approach zero, we should make $\Delta\varepsilon_1 = -\varepsilon_1$ and $\Delta\varepsilon_2 = -\varepsilon_2$

$$(-r_3 \sin\theta_3')\Delta\theta_3 + (-1)\Delta r_4 = -\varepsilon_1 \tag{11}$$

$$(r_3 \cos\theta_3')\Delta\theta_3 + (0)\Delta r_4 = -\varepsilon_2 \tag{12}$$

If equations (11) and (12) are solved for $\Delta\theta_3$ and Δr_4, then new and better estimates for θ_3 and r_4 will be

$$\theta_3' \text{ (new)} = \theta_3' \text{ (old)} + \Delta\theta_3 \tag{13}$$

$$r_4' \text{ (new)} = r_4' \text{ (old)} + \Delta r_4 \tag{14}$$

Numerical Example

Data $r_2 = 2$ in. $r_3 = 4$ in. $r_1 = 1$ in.

$\theta_2 = 60$ deg.

<u>First estimates</u>

θ'_3 = 330 deg. r'_4 = 5 in.

<u>Calculate errors</u>

(Equation (3)): ε_1 = 2 cos 60° + 4 cos 330° − 5 = −.540

(Equation (4)): ε_2 = 2 sin 60° + 4 sin 330° + 1 = +.732

<u>Substitute in equations (11) and (12) and solve</u>

(2.00) $\Delta\theta_3$ − Δr_4 = .540

(3.46) $\Delta\theta_3$ = −.732

$\Delta\theta_3$ = −.211 radians or −12.1 deg.

Δr_4 = −.963 in.

<u>New estimates</u>

θ'_3 = 330 − 12.1 = 317.9 deg.

r'_4 = 5 − .963 = 4.037 in.

<u>New errors</u>

ε_1 = −.069

ε_2 = .050

<u>New corrections</u> (Equations (11) and (12) again)

$\Delta\theta_3$ = −.017 rad   or   −.97 deg.

Δr_4 = −.114 in.

$$\theta_3' = 317.9 - .97 = 316.93 \text{ deg.}$$

$$r_4' = 4.037 - .114 = 3.923 \text{ in.}$$

If this process is continued until the corrections are no more than .01 deg. and .001 in. the results are:

$$\theta_3 = 316.92 \text{ deg.}$$

$$r_4 = 3.922 \text{ in.}$$

1.10 Position Solution, Closed Form, Slider-Crank Example

By a closed-form solution we mean the deriving of explicit relations for calculating the variable unknowns. Using the slider-crank mechanism (Fig. 1.11) as an example, we start with the vector loop equation.

$$\bar{r}_2 + \bar{r}_3 - \bar{r}_4 - \bar{r}_1 = 0 \tag{1}$$

The problem is first to eliminate one of the unknowns. We choose to eliminate θ_3 by the following operations.

Rewrite equation (1):

$$\bar{r}_3 = \bar{r}_4 + \bar{r}_1 - \bar{r}_2 \tag{2}$$

Do the following:

$$\bar{r}_3 \cdot \bar{r}_3 = (\bar{r}_4 + \bar{r}_1 - \bar{r}_2) \cdot (\bar{r}_4 + \bar{r}_1 - \bar{r}_2) \tag{3}$$

From which:

$$r_3^2 = r_4^2 + r_1^2 + r_2^2 + 2\overline{r}_4 \cdot \overline{r}_1 - 2\overline{r}_4 \cdot \overline{r}_2 - 2\overline{r}_1 \cdot \overline{r}_2 \tag{4}$$

$$r_3^2 = r_4^2 + r_1^2 + r_2^2 - 2r_4 r_2 \cos\theta_2 + 2r_1 r_2 \sin\theta_2 \tag{5}$$

Rearrange into standard "quadratic" form:

$$r_4^2 + (-2r_2\cos\theta_2)r_4 + (r_1^2 + r_2^2 - r_3^2 + 2r_1 r_2 \sin\theta_2) = 0 \tag{6}$$

<u>Solution for r_4</u>

$$r_4 = r_2\cos\theta_2 \pm \sqrt{r_2^2\cos^2\theta_2 - r_1^2 - r_2^2 + r_3^2 - 2r_1 r_2 \sin\theta_2} \tag{7}$$

<u>Solution for θ_3</u>

Loop equation in component form:

$$r_2\cos\theta_2 + r_3\cos\theta_3 - r_4 = 0 \tag{8}$$

$$r_2\sin\theta_2 + r_3\sin\theta_3 + r_1 = 0 \tag{9}$$

From which:

$$\cos\theta_3 = (r_4 - r_2\cos\theta_2)/r_3 \tag{10}$$

$$\sin\theta_3 = (-r_2\sin\theta_2 - r_1)/r_3 \tag{11}$$

Numerical Example

<u>Data</u> $r_2 = 2$ in. $r_3 = 4$ in. $r_1 = 1$ in. $\theta_2 = 60$ deg.

<u>After substitution in equation (7)</u>

$r_4 = 3.922$ in.   or -1.922 in.

1.17

From equations (10) and (11)

If $r_4 = 3.922$ in., then $\cos\theta_3 = .73050$

$$\sin\theta_3 = -.68301$$

$$\theta_3 = 316.92 \text{ deg.}$$

If $r_4 = -1.922$ in., then $\cos\theta_3 = -.73050$

$$\sin\theta_3 = -.68301$$

$$\theta_3 = 223.08 \text{ deg.}$$

The two solutions correspond to two different ways the mechanism could be assembled (Fig. 1.12). The intended solution is, of course, the one shown in solid lines. This situation arises in the solution of any vector loop equation. If one is doing a hand calculation for one position, it is no problem to decide which is the desired solution. If the problem is being programmed for automatic computation, then some serious thought must be given to the logic needed in the program to select the proper solution. The iterative solution illustrated earlier avoids this problem because it will ordinarily converge to the solution closer to the estimate.

1.11 Velocity and Acceleration Solutions

Our outline of the vector loop approach to kinematic analysis listed the final steps as follows.

(5) Differentiate the position equations with respect to time. Solve those for the velocity unknowns.

(6) Differentiate again with respect to time. Solve for the acceleration unknowns.

Once the position solution is complete, these steps are very straightforward. The velocity and acceleration equations are linear in the unknowns, so easily

1.18

solved. To illustrate, we will continue with a numerical example for the slider-crank (Fig. 1.11).

<u>Fixed data</u> $r_2 = 2$ in. $r_3 = 4$ in. $r_1 = 1$ in.

<u>Input position</u>

$\theta_2 = 60$ deg.

<u>From position solution</u>

$\theta_3 = 316.92$ deg. $r_4 = 3.922$ in.

The position equations were

$$r_2\cos\theta_2 + r_3\cos\theta_3 - r_4 = 0 \tag{1}$$

$$r_2\sin\theta_2 + r_3\sin\theta_3 + r_1 = 0 \tag{2}$$

Differentiation with respect to time yields the velocity equations

$$-r_2\dot{\theta}_2\sin\theta_2 - r_3\dot{\theta}_3\sin\theta_3 - \dot{r}_4 = 0 \tag{3}$$

$$r_2\dot{\theta}_2\cos\theta_2 + r_3\dot{\theta}_3\cos\theta_3 = 0 \tag{4}$$

Unknowns are $\dot{\theta}_3$ and $\dot{r}_4$

"Standard form" of equations (3) and (4) is:

$$(-r_3\sin\theta_3)\dot{\theta}_3 + (-1)\dot{r}_4 = r_2\dot{\theta}_2\sin\theta_2 \tag{5}$$

$$(r_3\cos\theta_3)\dot{\theta}_3 + (0)\dot{r}_4 = -r_2\dot{\theta}_2\cos\theta_2 \tag{6}$$

Substitution of known values in equations (5) and (6) yields

$$2.732\dot{\theta}_3 - \dot{r}_4 = 1.732\dot{\theta}_2$$

$$2.922\dot{\theta}_3 = -1.000\dot{\theta}_2$$

From which,

$$\dot{\theta}_3 = -.342\dot{\theta}_2$$

$$\dot{r}_4 = -2.666\dot{\theta}_2$$

Differentiation of the velocity equations with respect to time produces the acceleration equations.

$$-r_2\ddot{\theta}_2\sin\theta_2 - r_2\dot{\theta}_2^2\cos\theta_2 - r_3\ddot{\theta}_3\sin\theta_3 - r_3\dot{\theta}_3^2\cos\theta_3 - \ddot{r}_4 = 0 \qquad (7)$$

$$r_2\ddot{\theta}_2\cos\theta_2 - r_2\dot{\theta}_2^2\sin\theta_2 + r_3\ddot{\theta}_3\cos\theta_3 - r_3\dot{\theta}_3^2\sin\theta_3 = 0 \qquad (8)$$

Unknowns are $\ddot{\theta}_3$ and $\ddot{r}_4$

"Standard form" of equations (7) and (8) is:

$$(-r_3\sin\theta_3)\ddot{\theta}_3 - \ddot{r}_4 = r_2\ddot{\theta}_2\sin\theta_2 + r_2\dot{\theta}_2^2\cos\theta_2 + r_3\dot{\theta}_3^2\cos\theta_3 \qquad (9)$$

$$(r_3\cos\theta_3)\ddot{\theta}_3 + (0)\ddot{r}_4 = -r_2\ddot{\theta}_2\cos\theta_2 + r_2\dot{\theta}_2^2\sin\theta_2 + r_3\dot{\theta}_3^2\sin\theta_3 \qquad (10)$$

Substitution of the fixed data, plus the results from the position and velocity solutions, reduces these equations to

$$2.732\ddot{\theta}_3 - \ddot{r}_4 = 1.732\ddot{\theta}_2 + 1.342\dot{\theta}_2^2$$

$$2.922\ddot{\theta}_3 = -1.000\ddot{\theta}_2 + 1.412\dot{\theta}_2^2$$

from which,

$$\ddot{\theta}_3 = -.342\ddot{\theta}_2 + .483\dot{\theta}_2^2$$

$$\ddot{r}_4 = -2.666\ddot{\theta}_2 - 0.064\dot{\theta}_2^2$$

1.12 Motion of Points

The discussion so far has concentrated on the solution of the basic vector loops describing the motions of the links of a mechanism. Particular points (for example, the centers of mass for the different links) will be of special interest. We will need to compute position, velocity and acceleration for such points. These computations follow after the basic loop analysis, which must always be done first.

For an example, consider the 4-bar mechanism of Fig. 1.13. The basic vector loop position, velocity and acceleration equations are

$$\bar{r}_2 + \bar{r}_3 - \bar{r}_4 - \bar{r}_1 = 0 \tag{1}$$

$$\dot{\bar{r}}_2 + \dot{\bar{r}}_3 - \dot{\bar{r}}_4 = 0 \tag{2}$$

$$\ddot{\bar{r}}_2 + \ddot{\bar{r}}_3 - \ddot{\bar{r}}_4 = 0 \tag{3}$$

We assume that these have been solved, so that we now know all the following.

Given constants $\qquad r_1, \quad r_2, \quad r_3, \quad r_4$

Input position and state of motion

$$\theta_2, \quad \dot{\theta}_2, \quad \ddot{\theta}_2$$

Variables determined by solving equations (1), (2) and (3)

$$\theta_3, \quad \dot{\theta}_3, \quad \ddot{\theta}_3, \quad \theta_4, \quad \dot{\theta}_4, \quad \ddot{\theta}_4$$

We now wish to determine the state of motion for point "C" of link 3. We first describe its position in the X,Y system by the vector $\bar{r}_c$.

$$\bar{r}_c = \bar{r}_2 + \bar{r}_5 \tag{4}$$

where $\bar{r}_5$ is a new vector (not part of the basic loop for the mechanism) drawn on link 3. The length of $\bar{r}_5$ and the angle, λ, from vector $\bar{r}_3$ to $\bar{r}_5$ are known.

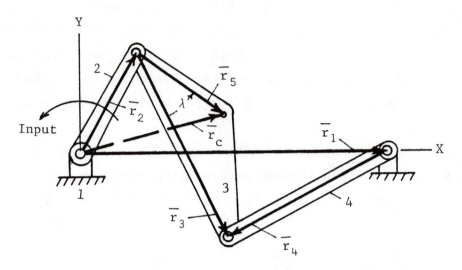

Fig. 1.13   4-bar mechanism.   Motion of point C to be determined.

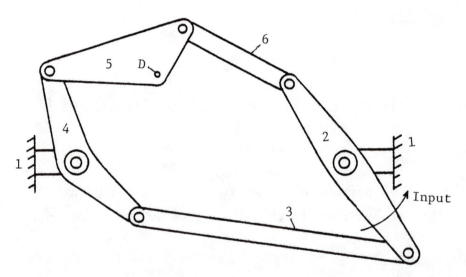

Fig. 1.14   A 6-link mechanism.   Motion of point D to be determined.

Therefore everything on the right side of equation (4) is known. We calculate the X and Y coordinates for the location of point C.

$$X_c = r_2\cos\theta_2 + r_5\cos(\theta_3 + \lambda) \tag{5}$$

$$Y_c = r_2\sin\theta_2 + r_5\sin(\theta_3 + \lambda) \tag{6}$$

The X and Y components for the velocity of point C are determined by differentiating equations (5) and (6) with respect to time.

$$\dot{X}_c = (-r_2\sin\theta_2)\dot{\theta}_2 - r_5\sin(\theta_3 + \lambda)\dot{\theta}_3 \tag{7}$$

$$\dot{Y}_c = (r_2\cos\theta_2)\dot{\theta}_2 + r_5\cos(\theta_3 + \lambda)\dot{\theta}_3 \tag{8}$$

The acceleration components are determined by differentiating again with respect to time.

$$\ddot{X}_c = (-r_2\sin\theta_2)\ddot{\theta}_2 - (r_2\cos\theta_2)\dot{\theta}_2^2 - r_5\sin(\theta_3 + \lambda)\ddot{\theta}_3$$
$$- r_5\cos(\theta_3 + \lambda)\dot{\theta}_3^2 \tag{9}$$

$$\ddot{Y}_c = (r_2\cos\theta_2)\ddot{\theta}_2 - (r_2\sin\theta_2)\dot{\theta}_2^2 + r_5\cos(\theta_3 + \lambda)\ddot{\theta}_3$$
$$- r_5\sin(\theta_3 + \lambda)\dot{\theta}_3^2 \tag{10}$$

1.13 Point Motion, General Remarks

Once the basic loop for a mechanism has been solved it will always be possible to describe the position of any point of any link as the sum of known vectors, which can then be differentiated to yield the velocity and acceleration of the point. Usually there will be more than one way to do this. Consider the mechanism of Fig. 1.14 and the problem of determining the motion of point D.

First we must define suitable vectors for the basic mechanism analysis. A possible choice is pictured in Fig. 1.15. Loop equations are

$$\bar{r}_2 + \bar{r}_3 + \bar{r}_4 + \bar{r}_1 = 0 \tag{1}$$

(Variable unknowns θ_3, θ_4)

$$\bar{r}_7 + \bar{r}_6 + \bar{r}_5 + \bar{r}_8 + \bar{r}_1 = 0 \tag{2}$$

(Variable unknowns θ_6, θ_5, θ_4)

(Note: $\theta_7 = \theta_2$ + constant and $\theta_8 = \theta_4$ + constant)

Assuming these have been solved, and the basic loop analysis completed by solving the corresponding velocity and acceleration loops, we will know everything needed to proceed with studying the motion of point D.

The position of point D might be described as shown in Fig. 1.16.

$$\bar{r}_d = \bar{r}_7 + \bar{r}_6 + \bar{r}_9 \tag{3}$$

The corresponding coordinate expressions would be

$$X_d = r_7 \cos\theta_7 + r_6 \cos\theta_6 + r_9 \cos(\theta_5 + \gamma) \tag{4}$$

$$Y_d = r_7 \sin\theta_7 + r_6 \sin\theta_6 + r_9 \sin(\theta_5 + \gamma) \tag{5}$$

An equally correct (and slightly shorter) description would be as shown in Fig. 1.17.

$$\bar{r}_d = -\bar{r}_1 - \bar{r}_8 + \bar{r}_{10} \tag{6}$$

$$X_d = -r_1 - r_8 \cos\theta_8 + r_{10} \cos(\theta_5 + \lambda) \tag{7}$$

$$Y_d = -r_8 \sin\theta_8 + r_{10} \sin(\theta_5 + \lambda) \tag{8}$$

It would also be quite correct (Fig. 1.18), but a longer description to say

$$\bar{r}_d = \bar{r}_2 + \bar{r}_3 + \bar{r}_4 - \bar{r}_8 + \bar{r}_{10} \tag{9}$$

1.24

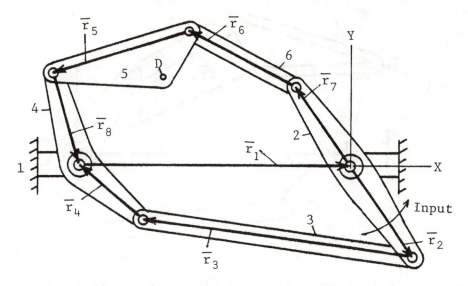

Fig. 1.15   Basic vector loops for the mechanism.

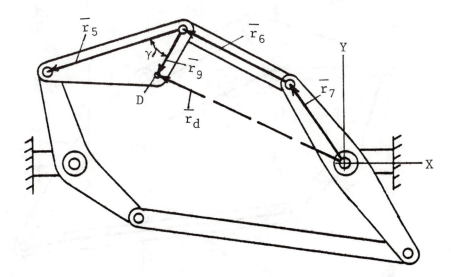

Fig. 1.16   One way to describe the position of D.

1.25

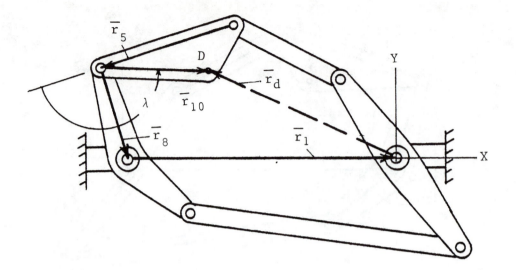

Fig. 1.17   Another description of the position vector for point D.

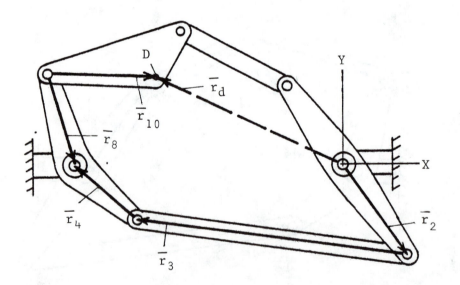

Fig. 1.18   A correct, but long, description of the position of point D.

1.14 Kinematic Coefficients, Definitions

The term "kinematic coefficient" will be used for derivatives of motion variables with respect to the "input" or reference variable.

Let S_i be the input variable (scalar)

S_k be any other variable (scalar)

The variables S_i and S_k can be either angles or vector lengths. One might be an angle and the other a vector length. If S_k is an angle θ_k, then we will use the following symbolism.

$$d\theta_k/dS_i = h_k \tag{1}$$

$$d^2\theta_k/dS_i^2 = h_k' \tag{2}$$

If S_k is a vector length, r_k, we will use

$$dr_k/dS_i = f_k \tag{3}$$

$$d^2r_k/dS_i^2 = f_k' \tag{4}$$

Take notice of the dimensions involved. If both variables are angles, the dimensions are

$$h_k, \quad \text{dimensionless} \quad = d\theta_k/d\theta_i \tag{5}$$

$$h_k', \quad \text{dimensionless} \quad = d^2\theta_k/d\theta_i^2 \tag{6}$$

If both variables are vector lengths,

$$f_k, \quad \text{dimensionless} \quad = dr_k/dr_i \tag{7}$$

$$f_k', \quad (\text{length})^{-1} \quad = d^2r_k/dr_i^2 \tag{8}$$

1.27

If the input variable is an angle θ_i, but the variable of interest is a vector length, r_k,

$$f_k, \quad \text{length} \quad = dr_k/d\theta_i \tag{9}$$

$$f_k', \quad \text{length} \quad = d^2r_k/d\theta_i^2 \tag{10}$$

If the input variable is a vector length r_i, but the variable of interest is an angle θ_k

$$h_k, \quad (\text{length})^{-1} \quad = d\theta_k/dr_i \tag{11}$$

$$h_k', \quad (\text{length})^{-2} \quad = d^2\theta_k/dr_i^2 \tag{12}$$

1.15 Kinematic Coefficients, Meaning

By the chain rule for differentiation,

$$dS_k/dt = (dS_k/dS_i)(dS_i/dt)$$

or
$$\dot{S}_k = (dS_k/dS_i)\dot{S}_i \tag{1}$$

and
$$d^2S_k/dt^2 = (dS_k/dS_i)(d^2S_i/dt^2) + (d^2S_k/dS_i^2)(dS_i/dt)^2$$

or
$$\ddot{S}_k = (dS_k/dS_i)\ddot{S}_i + (d^2S_k/dS_i^2)\dot{S}_i^2 \tag{2}$$

Hence, depending upon whether the variable of interest is an angle θ_k or a vector length r_k, we can say

$$\dot{\theta}_k = h_k\dot{S}_i \tag{3}$$

$$\ddot{\theta}_k = h_k\ddot{S}_i + h_k'\dot{S}_i^2 \tag{4}$$

$$\dot{r}_k = f_k\dot{S}_i \tag{5}$$

$$\ddot{r}_k = f_k\ddot{S}_i + f_k'\dot{S}_i^2 \tag{6}$$

The significance of the kinematic coefficients lies in the fact that they are functions of _position_ only. The numerical values are _constant for a fixed position of the input link_; they do not depend upon the state of motion (velocity, acceleration) of the input.

For an example, consider the slider-crank problem solved earlier (Fig. 1.11). The position equations were

$$r_2\cos\theta_2 + r_3\cos\theta_3 - r_4 = 0 \tag{7}$$

$$r_2\sin\theta_2 + r_3\sin\theta_3 + r_1 = 0 \tag{8}$$

After differentiating with respect to the input variable, θ_2, we have

$$-r_2\sin\theta_2 - (r_3\sin\theta_3)h_3 - f_4 = 0 \tag{9}$$

$$r_2\cos\theta_2 + (r_3\cos\theta_3)h_3 = 0 \tag{10}$$

where, $\quad h_3 = d\theta_3/d\theta_2$

$$f_4 = dr_4/d\theta_2$$

For r_2 = 2 in., r_3 = 4 in., r_1 = 1 in., at θ_2 = 60 deg. the solution to the position equations was

$$\theta_3 = 316.92 \text{ deg.} \qquad r_4 = 3.922 \text{ in.}$$

Using these results in equations (9) and (10) and solving for h_3 and f_4, the result is

$$h_3 = -0.342$$

$$f_4 = -2.666 \text{ in.}$$

Upon differentiating equations (9) and (10) with respect to θ_2 we have

1.29

$$-r_2\cos\theta_2 - (r_3\sin\theta_3)h_3' - (r_3\cos\theta_3)h_3^2 - f_4' = 0 \qquad (11)$$

$$-r_2\sin\theta_2 + (r_3\cos\theta_3)h_3' - (r_3\sin\theta_3)h_3^2 = 0 \qquad (12)$$

Using the previous results and solving for h_3' and f_4',

$$h_3' = 0.483$$

$$f_4' = -0.064 \text{ in.}$$

Equations (3)-(6) tell us

$$\dot{\theta}_3 = h_3\dot{\theta}_2$$

$$\ddot{\theta}_3 = h_3\ddot{\theta}_2 + h_3'\dot{\theta}_2^2$$

$$\dot{r}_4 = f_4\dot{\theta}_2$$

$$\ddot{r}_4 = f_4\ddot{\theta}_2 + f_4'\dot{\theta}_2^2 \quad .$$

Hence, for our numerical example,

$$\dot{\theta}_3 = -0.342\dot{\theta}_2$$

$$\ddot{\theta}_3 = -0.342\ddot{\theta}_2 + 0.483\dot{\theta}_2^2$$

$$\dot{r}_4 = -2.666\dot{\theta}_2$$

$$\ddot{r}_4 = -2.666\ddot{\theta}_2 - 0.064\dot{\theta}_2^2$$

These are the same results obtained in our earlier solution.

1.16 Example, Use of Kinematic Coefficients (Fig. 1.19)

We repeat. The kinematic coefficients are functions of position only. To see part of what this means, consider the following problem.

Given: We have a mechanism sealed in a housing (Fig. 1.19). The details of the mechanism itself are not known to us, but we have been able to experiment by moving, and making motion measurements on, three links which project from the housing. We have found that if we turn the shaft of link 2 counterclockwise at a constant angular velocity of 20 rad/sec, then, at the position illustrated, the shaft of link 4 and point C of link 3 have the states of motion indicated.

Question: If we were to turn the shaft of 2 clockwise at 10 rad/sec and also give it a clockwise angular acceleration of 150 rad/sec², what would be the resulting states of motion for link 4 and for point C, in the same mechanism position?

We can solve the problem as follows.

(1) Let the position, θ_2, of link 2, be the "input," or reference, variable. Then

$$h_4 = d\theta_4/d\theta_2$$

$$h_4' = d^2\theta_4/d\theta_2^2$$

$$f_{xc} = dX_c/d\theta_2$$

$$f_{xc}' = d^2X_c/d\theta_2^2$$

$$f_{yc} = dY_c/d\theta_2$$

$$f_{yc}' = d^2Y_c/d\theta_2^2 \ .$$

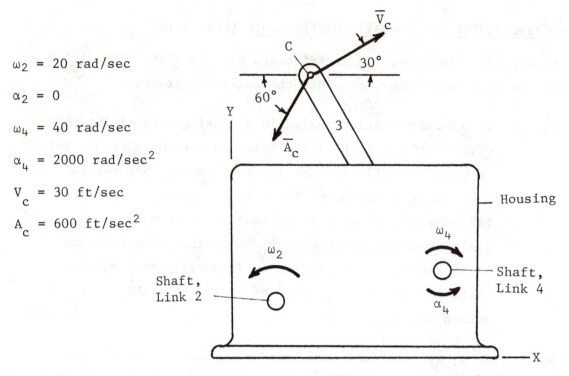

$\omega_2 = 20$ rad/sec

$\alpha_2 = 0$

$\omega_4 = 40$ rad/sec

$\alpha_4 = 2000$ rad/sec^2

$V_c = 30$ ft/sec

$A_c = 600$ ft/sec^2

Fig. 1.19 Example in use of kinematic coefficients. Initial data.

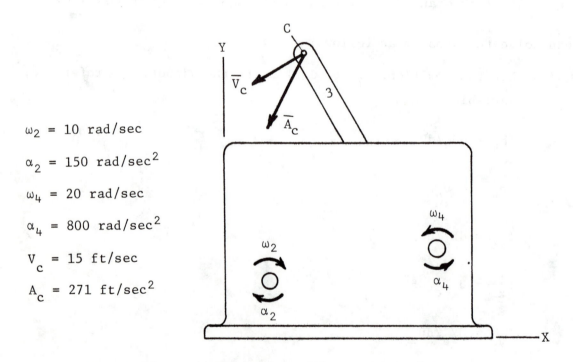

$\omega_2 = 10$ rad/sec

$\alpha_2 = 150$ rad/sec^2

$\omega_4 = 20$ rad/sec

$\alpha_4 = 800$ rad/sec^2

$V_c = 15$ ft/sec

$A_c = 271$ ft/sec^2

Fig. 1.20 State of motion with new input data.

Also, $\omega_4 = h_4 \omega_2$

$\alpha_4 = h_4 \alpha_2 + h_4' \omega_2^2$

$V_{xc} = f_{xc} \omega_2$

$A_{xc} = f_{xc} \alpha_2 + f_{xc}' \omega_2^2$

$V_{yc} = f_{yc} \omega_2$

$A_{yc} = f_{yc} \alpha_2 + f_{yc}' \omega_2^2$

(2) Use the given state-of-motion data to calculate all kinematic coefficients in the above equations.

$h_4 = \omega_4/\omega_2 = -40/20 = -2.00$

$h_4' = (\alpha_4 - h_4 \alpha_2)/\omega_2^2$

$= [2000 - (-2.00)(0)]/(20)^2 = 5.00$

$f_{xc} = V_{xc}/\omega_2 = (30\cos 30°)/20 = 1.30$ ft.

$f_{yc} = V_{yc}/\omega_2 = (30\sin 30°)/20 = 0.75$ ft.

$f_{xc}' = (A_{xc} - f_{xc} \alpha_2)/\omega_2^2$

$= [-600\cos 60° - (1.30)(0)]/(20)^2 = -0.75$ ft.

$f_{yc}' = (A_{yc} - f_{yc} \alpha_2)/\omega_2^2$

$= [-600\sin 60° - (0.75)(0)]/(20)^2 = -1.30$ ft.

1.33

(3) Use the new data concerning the state-of-motion of link 2 to calculate the new states-of-motion for link 4 and point C.

$$\omega_4 = h_4\omega_2 = (-2.00)(-10) = 20 \text{ rad/sec}$$

$$\alpha_4 = h_4\alpha_2 + h_4'\omega_2^2$$

$$= (-2.00)(-150) + (5.00)(-10)^2 = 800 \text{ rad/sec}^2$$

$$V_{xc} = f_{xc}\omega_2 = (1.30)(-10) = -13.0 \text{ ft/sec}$$

$$V_{yc} = f_{yc}\omega_2 = (0.75)(-10) = -7.5 \text{ ft/sec}$$

$$A_{xc} = f_{xc}\alpha_2 + f_{xc}'\omega_2^2$$

$$= (1.30)(-150) + (-0.75)(-10)^2 = -120 \text{ ft/sec}^2$$

$$A_{yc} = f_{yc}\alpha_2 + f_{yc}'\omega_2^2$$

$$= (0.75)(-150) + (-1.30)(-10)^2 = -243 \text{ ft/sec}^2$$

(4) Determine magnitude and direction of vectors $\overline{V}_c$ and $\overline{A}_c$.

$$V_c = \sqrt{V_{xc}^2 + V_{yc}^2} = \sqrt{(-13.0)^2 + (-7.5)^2} = 15.0 \text{ ft/sec}$$

Angle = arc tan $(-7.5/-13.0)$ = 210 deg.

$$A_c = \sqrt{A_{xc}^2 + A_{yc}^2} = \sqrt{(-120)^2 + (-243)^2} = 271 \text{ ft/sec}^2$$

Angle = arc tan $(-243/-120)$ = 243.7 deg.

The resulting situation is as shown in Fig. 1.20.

We chose to use θ_2 as the reference variable. This was logical, but not necessary. Suppose we had used θ_4 instead. Then we would write, for example,

$$h_2 = d\theta_2/d\theta_4$$

$$h_2' = d^2\theta_2/d\theta_4^2$$

and

$$\omega_2 = h_2\omega_4$$

$$\alpha_2 = h_2\alpha_4 + h_2'\omega_4^2$$

We would calculate the coefficients, using the initial data,

$$h_2 = \omega_2/\omega_4 = 20/-40 = -0.50$$

$$h_2' = (\alpha_2 - h_2\alpha_4)/\omega_4^2$$

$$= [0 - (-0.50)(2000)]/(-40)^2 = 0.625$$

Then, for the new data,

$$\omega_4 = \omega_2/h_2 = (-10)/-0.50 = 20 \text{ rad/sec}$$

$$\alpha_4 = (\alpha_2 - h_2'\omega_4^2)/h_2$$

$$= [(-150) - (0.625)(20)^2]/-0.50 = 800 \text{ rad/sec}^2$$

E1.1 For each mechanism shown define vector loops suitable for a complete
      kinematic analysis. Assume the motion of link 2 to be the known
      input. Write the vector loop position equations. List the variable
      scalar unknowns.

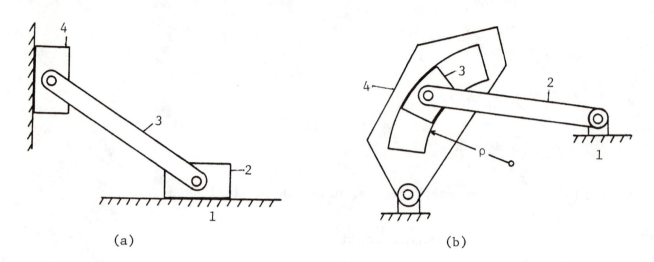

(a)

(b)

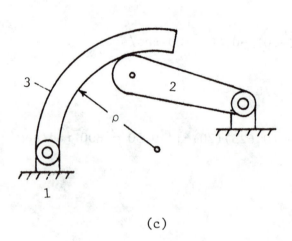

(c)

E1.2 For each mechanism shown define vector loops suitable for a complete
kinematic analysis. Assume the motion of link 2 to be the known
input. Write the vector loop position equations. List the variable
scalar unknowns.

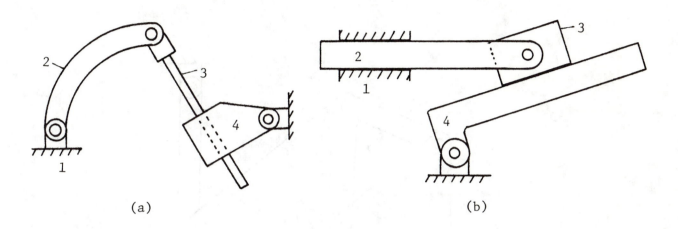

(a)   (b)

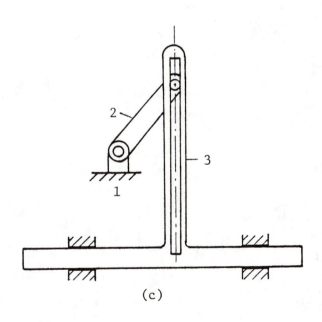

(c)

E1.3 For the mechanism shown define vector loops suitable for a complete
      kinematic analysis. Assume the motion of link 2 to be the known
      input. Write the vector loop position equations. List the variable
      scalar unknowns.

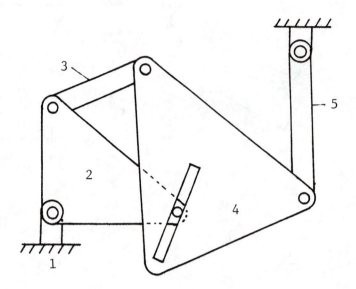

E1.4 For the mechanism shown define vector loops suitable for a complete
      kinematic analysis. Assume the motion of link 2 to be the known
      input. Write the vector loop position equations. List the variable
      scalar unknowns.

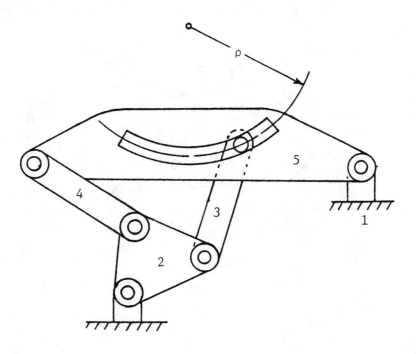

E1.5 For the mechanism shown define vector loops suitable for a complete
      kinematic analysis. Assume the motion of link 2 to be the known
      input. Write the vector loop position equations. List the variable
      scalar unknowns. (Note: The contacting portions of links 2 and 3
      are circular, with centers at C_2 and C_3 respectively.)

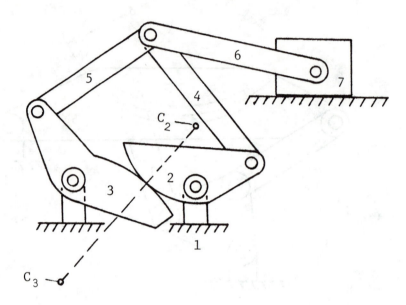

E1.6 Constants: $r_1 = 1$ in., $r_3 = 2$ in., $\Theta_1 = 90°$

Variables: r_2, r_4, Θ_4 ($\Theta_3 = \Theta_4 - 90°$)

Let $r_2 = 3$ in. and the velocity of link 2 be 20ft/sec to the right, constant. Determine the position variables r_4 and Θ_4 and their first two time derivatives.

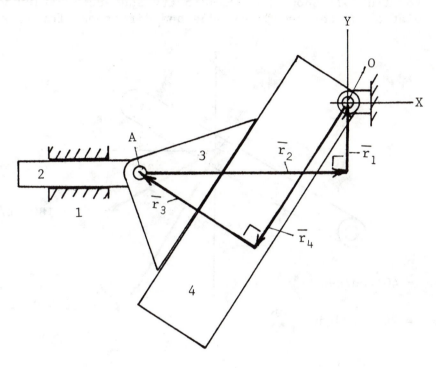

E1.7 Link 5 of a particular mechanism is moving instantaneously, as indicated below, when the driving crank is in a certain position and rotating clockwise at 120 rad/sec, constant.

For the same position of the mechanism, if the driving crank were rotating clockwise at 60 rad/sec and had an angular acceleration of 1800 rad/sec^2 counter-clockwise, what would be the state of motion for link 5? Show the new velocity and acceleration vectors for point G and the new magnitudes and directions for ω_5 and α_5.

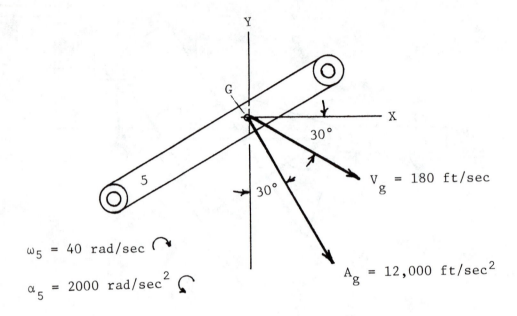

ω_5 = 40 rad/sec

α_5 = 2000 rad/sec^2

V_g = 180 ft/sec

A_g = 12,000 ft/sec^2

1.42

E1.8 Write the scalar position equations and, for θ_2 = 60 deg., solve by
the iterative method. Take as initial estimates θ_3 = 10 deg. and
θ_4 = 50 deg. (Since the linkage is a parallelogram the correct values
are obviously 0 deg. and 60 deg. However, the purpose of this exer-
cise is to test the iterative procedure.) Continue the iterations
until you are within 0.05 deg. of the correct answers.

OA = MB = 2 in.

OM = AB = 4 in.

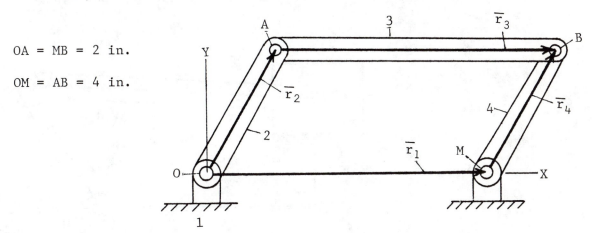

E1.9 The position equations have been solved. Θ_3 = 183.33 deg. and r_4 = 2.993 ft.

Write the position equations, differentiate and solve for the kinematic coefficients. Use these to determine the angular velocity and angular acceleration of link 3 and the velocity and acceleration of slider 4. Assume crank 2 to be rotating 40 rad/sec, clockwise and constant.

r_2 = 2.0 ft.

r_3 = 4.0 ft.

r_1 = 1.5 ft.

Θ_2 = 60 deg.

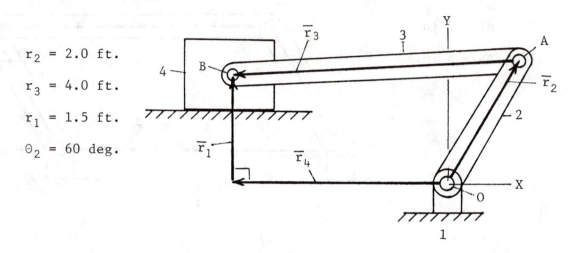

CHAPTER 2

KINEMATIC ANALYSIS II

2.1 Introduction

In the previous chapter we introduced the vector-loop approach to the
kinematic analysis of mechanisms, displayed a few examples, discussed briefly
the kinematics of point motion, and introduced the concept of kinematic coefficients. In this chapter we will elaborate somewhat on these ideas and show
some additional examples.

2.2 Rolling Contact, Circular Bodies

Previous examples did not include mechanisms in which a pair of bodies
(links) are of circular form and in rolling contact with each other. This is
not an uncommon situation in machines, so should be examined. Figure 2.1
illustrates two possible mechanisms involving this rolling-contact feature.
The "rolling" objects may actually have smooth surfaces maintained in rolling
(not slipping) by friction, or they may be toothed gears represented by their
rolling pitch circles. Thus, in Fig. 2.1(b), link 5 might be a toothed gear
in mesh with a gear forming a part of link 3.

The vector loops shown in Fig. 2.1 are needed, but are not sufficient to
fully describe the motion. For the mechanism of Fig. 2.1(a)

$$\bar{r}_2 + \bar{r}_3 + \bar{r}_4 - \bar{r}_5 + \bar{r}_1 = 0 \tag{1}$$

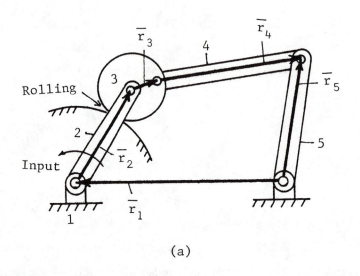

(a)

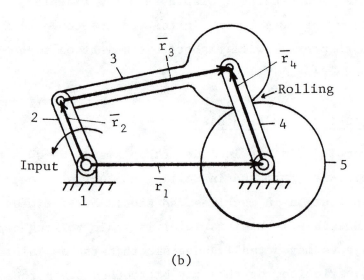

(b)

Fig. 2.1 Two mechanisms with rolling contact.

2.2

Variable unknowns are θ_3, θ_4, and θ_5. (θ_2 is the known input position.) This is one too many unknowns. The equation cannot be solved by itself.

For the mechanism of Fig. 2.1(b)

$$\bar{r}_2 + \bar{r}_3 - \bar{r}_4 - \bar{r}_1 = 0 \qquad\qquad (2)$$

Variable unknowns are θ_3 and θ_4. The equation can be solved, but does not tell us θ_5 (i.e. does not say anything about how link 5 moves).

In each of these examples we need one more <u>scalar</u> equation relating some of the variables of interest.

2.3 The Assembly Position

It must first be recognized that no solution can be made to the problems posed in the previous article unless we know how the machine was <u>assembled</u> initially. In the mechanism of Fig. 2.1(a), for example, we must know, for some value of θ_2, the corresponding value of θ_3. These known values will be designated θ_{2i} and θ_{3i}. Suppose we assemble the mechanism as shown by solid lines in Fig. 2.2(a) ($\theta_{3i} = 270°$ at $\theta_{2i} = 0°$). After the crank has turned to a new position, $\theta_2 = 90°$, the mechanism configuration will be as shown in broken lines. If, however, we assemble as shown in Fig. 2.2(b) ($\theta_{3i} = 0°$ at $\theta_{2i} = 0$), then, after the crank has turned to $\theta_2 = 90°$, the mechanism configuration will be quite different.

2.4 Rolling Contact Equations

In Fig. 2.3(a) we have shown two circular objects in rolling contact drawn in the assembly position. Vector $\bar{r}_{4i}$ is the vector between circle centers, $\bar{r}_{3i}$ is drawn on body 3 and $\bar{r}_{2i}$ is drawn on body 2. The two rolling bodies are in contact at point P.

The two bodies are shown again in Fig. 2.3(b) after some plane motion has taken place. Nothing is assumed stationary. The body centers have moved and some turning has taken place. The bodies are now in contact at point Q. The points P_2 and P_3 which were originally in contact at P are now located at P_2' and P_3', respectively. Body 2 (hence vector $\bar{r}_2$) has turned through an angle we will call $\Delta\theta_2$, body 3 (hence vector $\bar{r}_3$) has turned $\Delta\theta_3$ and vector $\bar{r}_4$ has turned

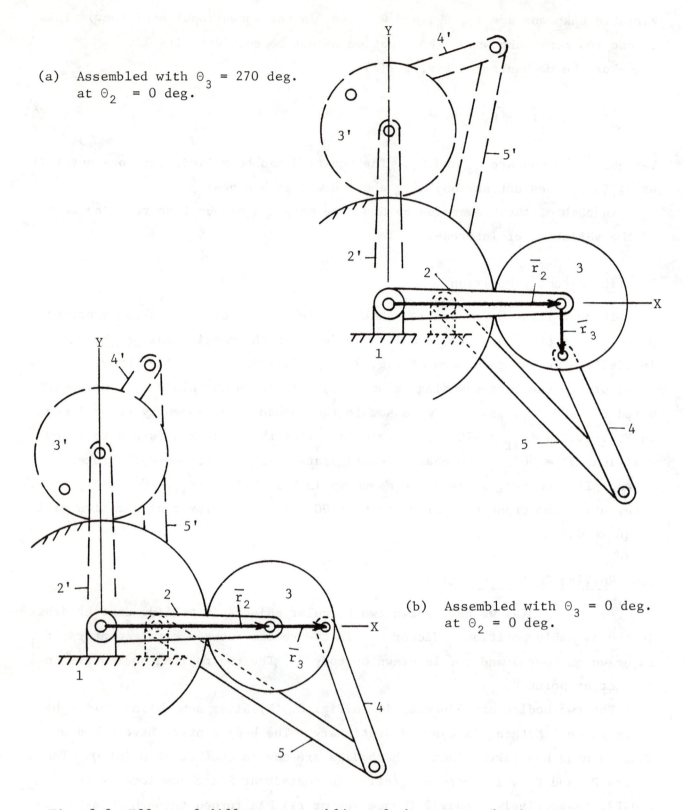

(a) Assembled with $\Theta_3 = 270$ deg. at $\Theta_2 = 0$ deg.

(b) Assembled with $\Theta_3 = 0$ deg. at $\Theta_2 = 0$ deg.

Fig. 2.2 Effect of different assemblies of the same mechanism parts.

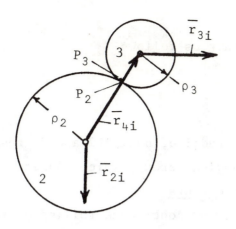
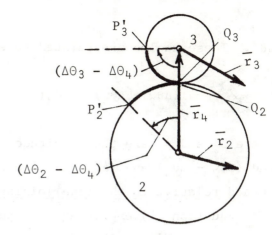

(a) Initial position.   (b)   Later position.

Fig. 2.3 Rolling contact.

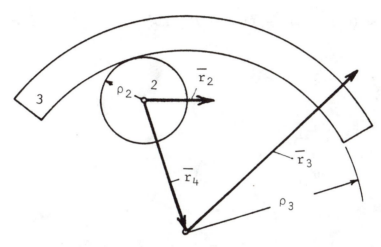

Fig. 2.4 Rolling. $\Delta\Theta_2 - (\rho_3/\rho_2)\Delta\Theta_3 - (1 - \rho_3/\rho_2)\Delta\Theta_4 = 0$.

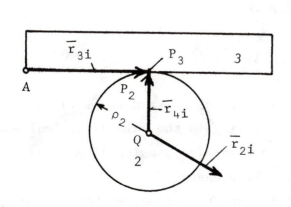
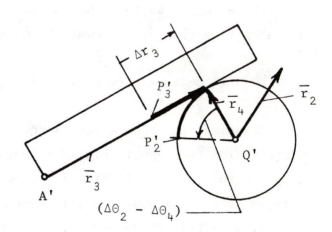

(a)   Initial position.   (b)   Later position.

Fig. 2.5 Rolling.

2.5

$\Delta\theta_4$. These angles are related to each other as follows

$$\rho_2(\Delta\theta_2 - \Delta\theta_4) = -\rho_3(\Delta\theta_3 - \Delta\theta_4) \qquad (1)$$

where ρ_2 and ρ_3 are the magnitudes of the radii of bodies 2 and 3, respectively, and $(\Delta\theta_2 - \Delta\theta_4)$ and $(\Delta\theta_3 - \Delta\theta_4)$ are the angles through which bodies 2 and 3 have turned <u>relative</u> <u>to</u> <u>the</u> <u>line</u> <u>joining</u> <u>their</u> <u>centers</u>.

Equation (1) expresses the fact that, for continuous rolling contact, the arc lengths QP_3' and QP_2' which have passed through contact must be equal. Upon a slight rearrangement this equation becomes

$$\Delta\theta_2 + (\rho_3/\rho_2)\Delta\theta_3 - (1 + \rho_3/\rho_2)\Delta\theta_4 = 0 \qquad (2)$$

Since

$$\Delta\theta_2 = \theta_2 - \theta_{2i}$$

$$\Delta\theta_3 = \theta_3 - \theta_{3i}$$

and

$$\Delta\theta_4 = \theta_4 - \theta_{4i}$$

we have

$$(\theta_2 - \theta_{2i}) + (\rho_3/\rho_2)(\theta_3 - \theta_{3i}) - (1 + \rho_3/\rho_2)(\theta_4 - \theta_{4i}) = 0 \qquad (3)$$

It is suggested that the reader study the situation represented in Fig. 2.4 and show that the rolling contact equation is

$$(\theta_2 - \theta_{2i}) - (\rho_3/\rho_2)(\theta_3 - \theta_{3i}) - (1 - \rho_3/\rho_2)(\theta_4 - \theta_{4i}) = 0 \qquad (4)$$

Sometimes the convention is adopted of setting the numerical value of ρ_3/ρ_2 positive for the case in Fig. 2.4 (bodies curving in the same direction) and negative for the case in Fig. 2.3 (bodies curving in opposite directions). If this is done, then equation (4) serves for both cases.

If we differentiate equation (4) with respect to time, we have the corresponding velocity equation,

$$\dot{\theta}_2 - (\rho_3/\rho_2)\dot{\theta}_3 - (1 - \rho_3/\rho_2)\dot{\theta}_4 = 0 \qquad (5)$$

Differentiating again, we have the acceleration equation,

$$\ddot{\theta}_2 - (\rho_3/\rho_2)\ddot{\theta}_3 - (1 - \rho_3/\rho_2)\ddot{\theta}_4 = 0 \qquad (6)$$

2.5 Special Case, One Radius Infinite

Figure 2.5 shows two bodies, one of which has an infinite radius, in rolling contact. It is inconvenient in this case to deal with a vector between the centers of curvature, so we define vectors as shown in the figure.

$\bar{r}_3$, drawn on 3, tail fixed on 3 at A, head always at point of contact, variables r_3 and θ_3.

$\bar{r}_4$, drawn from center of 2 to the point of contact (always perpendicular to $\bar{r}_3$). $\theta_4 = \theta_3 + 90$ deg.

$\bar{r}_2$, any vector drawn on 2.

After motion from the assembly position, Fig. 2.5(a), the configuration is as shown in Fig. 2.5(b). Both bodies have moved. Points P_3 and P_2, originally coincident, are now at P_3' and P_2'. The arc lengths QP_2' and QP_3' must be equal. Hence

$$\rho_2(\Delta\theta_2 - \Delta\theta_4) = \Delta r_3 \qquad (1)$$

or

$$(\theta_2 - \theta_{2i}) - (\theta_4 - \theta_{4i}) - (r_3 - r_{3i})/\rho_2 = 0 \qquad (2)$$

Differentiation with respect to time yields the velocity and acceleration equations

$$\dot{\theta}_2 - \dot{\theta}_4 - \dot{r}_3/\rho_2 = 0 \qquad (3)$$

$$\ddot{\theta}_2 - \ddot{\theta}_4 - \ddot{r}_3/\rho_2 = 0 \qquad (4)$$

2.7

2.6 Example Problem, Position Solution (Fig. 2.6)

The mechanism consists of input gear 2 in mesh with gear 4, link 3 pinned between gear centers and link 5 pinned to 4 and to the fixed link. The drawing shows what we will call the <u>assembly</u> position. Points O, A and B are in line, which puts link 5 in its extreme clockwise position. Link lengths and gear radii are listed on the drawing.

We wish to set up the computations to determine the motions of 3, 4, and 5 versus the input motion.

Problem set-up

(1) Form the indicated vector loop consisting of $\bar{r}_3$, $\bar{r}_4$, $\bar{r}_5$ and $\bar{r}_1$. Choose a coordinate system. Define arbitrarily a vector $\bar{r}_2$, drawn on gear 2. We choose to define $\bar{r}_2$ so that, in the assembly position, $\theta_2 = 0$.

(2) Write the vector loop equation and break into the X and Y scalar position equations.

$$\bar{r}_3 + \bar{r}_4 - \bar{r}_5 - \bar{r}_1 = 0 \tag{1}$$

$$r_3\cos\theta_3 + r_4\cos\theta_4 - r_5\cos\theta_5 - r_1 = 0 \tag{2}$$

$$r_3\sin\theta_3 + r_4\sin\theta_4 - r_5\sin\theta_5 = 0 \tag{3}$$

(3) In the assembly position $\theta_3 = \theta_4 = \theta_{3i}$, hence

$$(r_3 + r_4)\cos\theta_{3i} - r_5\cos\theta_{5i} - r_1 = 0$$

$$(r_3 + r_4)\sin\theta_{3i} - r_5\sin\theta_{5i} = 0$$

$$r_3 = 6\text{in.}, \quad r_4 = 1\text{in.}, \quad r_5 = 7\text{in.}, \quad r_1 = 4\text{in.}$$

$$7\cos\theta_{3i} - 7\cos\theta_{5i} - 4 = 0$$

$$7\sin\theta_{3i} - 7\sin\theta_{5i} = 0$$

$$\sin\theta_{5i} = \sin\theta_{3i}$$

So
$$\theta_{5i} = \theta_{3i} \quad \text{or} \quad 180 - \theta_{3i}$$

2.8

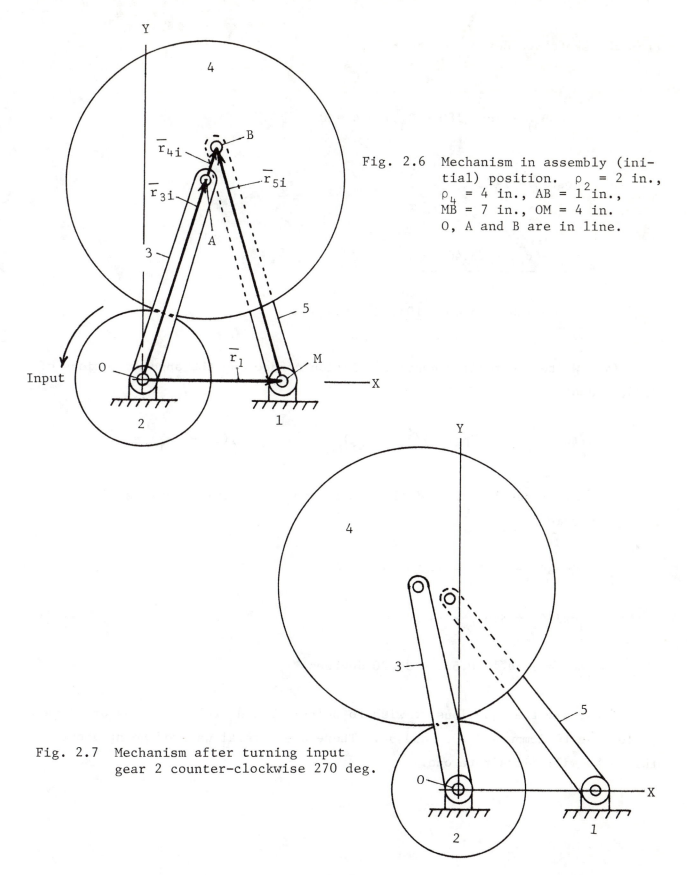

Fig. 2.6 Mechanism in assembly (initial) position. ρ_2 = 2 in., ρ_4 = 4 in., AB = 1 in., MB = 7 in., OM = 4 in. O, A and B are in line.

Fig. 2.7 Mechanism after turning input gear 2 counter-clockwise 270 deg.

2.9

From the sketch we see $\theta_{5i} = 180 - \theta_{3i}$

Hence

$$7\cos\theta_{3i} - 7\cos(180 - \theta_{3i}) - 4 = 0$$

or

$$\cos\theta_{3i} = 2/7,$$

so

$$\underline{\theta_{3i} = \theta_{4i} = 73.40 \text{ deg.}}$$

$$\underline{\theta_{5i}} = 180 - 73.40 = \underline{106.60} \text{ deg.}$$

(4) Write the rolling contact relation connecting the angular motions of 2, 3, and 4

$$(\theta_2 - \theta_{2i}) + (\rho_4/\rho_2)(\theta_4 - \theta_{4i}) - (1 + \rho_4/\rho_2)(\theta_3 - \theta_{3i}) = 0 \qquad (4)$$

(5) After substitution of the fixed numerical data, equations (2), (3) and (4) become

$$6\cos\theta_3 + \cos\theta_4 - 7\cos\theta_5 - 4 = 0 \qquad (5)$$

$$6\sin\theta_3 + \sin\theta_4 - 7\sin\theta_5 = 0 \qquad (6)$$

$$\theta_4 = 1.5\theta_3 - 0.5\theta_2 - 36.70 \text{ degrees} \qquad (7)$$

These are the equations we wish to solve. For different values of θ_2 we would like to compute θ_3, θ_4 and θ_5. There are several ways of going about this. We will illustrate one.

Solution

(1) Take the assembly position to be the "initial" position for starting the computation. In this position we know

$$\theta_2 = 0 \text{ deg}, \quad \theta_3 = 73.40 \text{ deg}, \quad \theta_4 = 73.40 \text{ deg}, \quad \theta_5 = 106.60 \text{ deg}.$$

(2) Choose a step size for θ_2. We will try 30 deg. At $\theta_2 = 30°$, estimate θ_3, θ_4 and θ_5. If not too large a step is taken, we can use initial values as good enough estimates of θ_3 and θ_5. An estimate of θ_4 can be made using equation (7).

$$\theta_3' = 73.40 \text{ deg}. \qquad \theta_5' = 106.60 \text{ deg}.$$

$$\theta_4' = 1.5(73.40) - .5(30) - 36.70 = 58.40 \text{ deg}.$$

(3) We will plan an iterative solution starting with the above estimates. First we define "errors"

$$\varepsilon_1 = 6\cos\theta_3' + \cos\theta_4' - 7\cos\theta_5' - 4 \tag{8}$$

$$\varepsilon_2 = 6\sin\theta_3' + \sin\theta_4' - 7\sin\theta_5' \tag{9}$$

$$\partial\varepsilon_1/\partial\theta_3' = -6\sin\theta_3' - \sin\theta_4'(\partial\theta_4'/\partial\theta_3')$$

$$= -6\sin\theta_3' - 1.5\sin\theta_4' = a_1 \tag{10}$$

$$\partial\varepsilon_1/\partial\theta_5' = 7\sin\theta_5' \qquad\qquad = b_1 \tag{11}$$

$$\partial\varepsilon_2/\partial\theta_3' = 6\cos\theta_3' + \cos\theta_4'(\partial\theta_4'/\partial\theta_3')$$

$$= 6\cos\theta_3' + 1.5\cos\theta_4' = a_2 \tag{12}$$

$$\partial\varepsilon_2/\partial\theta_5' = -7\cos\theta_5' \qquad\qquad = b_2 \tag{13}$$

We will calculate corrections $\Delta\theta_3$ and $\Delta\theta_5$ using the approximations

$$a_1\Delta\theta_3 + b_1\Delta\theta_5 = \Delta\varepsilon_1 = -\varepsilon_1 \qquad (14)$$

$$a_2\Delta\theta_3 + b_2\Delta\theta_5 = \Delta\varepsilon_2 = -\varepsilon_2 \qquad (15)$$

(4) $\varepsilon_1 = \cos 58.40° + 6\cos 73.40° - 7\cos 106.60° - 4 = 0.23793$

$\varepsilon_2 = \sin 58.40° + 6\sin 73.40° - 7\sin 106.60° \qquad = -0.10660$

$a_1 = -6\sin 73.40° - 1.5\sin 58.40° = -7.0275$

$b_1 = 7\sin 106.60° \qquad\qquad\qquad = 6.7083$

$a_2 = 6\cos 73.40° + 1.5\cos 58.40° \ = 2.5001$

$b_2 = -7\cos 106.60° \qquad\qquad\quad = 1.9998$

We solve

$$-7.0275\Delta\theta_3 + 6.7083\Delta\theta_5 = -0.23793$$

$$2.5001\Delta\theta_3 + 1.9998\Delta\theta_5 = \ \ 0.10660$$

The results are

$$\Delta\theta_3 = 0.03863 \text{ rad } (2.21 \text{ deg.})$$

$$\Delta\theta_5 = 0.00501 \text{ rad } (0.29 \text{ deg.})$$

Improved estimates are

$$\theta_3' = 73.40 + 2.21 = 75.61 \text{ deg.}$$

$$\theta_5' = 106.60 + 0.29 = 106.89 \text{ deg.}$$

$$\theta_4' = 1.5(75.61) - .5(30) - 36.70 = 61.72 \text{ deg.}$$

(5)   Using these improved estimates, we repeat the above calculations, with the results

$$\varepsilon_1 = -0.00135 \qquad\qquad \varepsilon_2 = -0.00565$$

$$a_1 = -7.1327 \qquad\qquad b_1 = 6.6981$$

$$a_2 = 2.2018 \qquad\qquad b_2 = 2.0337$$

$$\Delta\theta_3 = 0.07 \text{ deg.} \qquad\qquad \Delta\theta_5 = 0.08 \text{ deg.}$$

More improved estimates then are

$$\theta_3' = 75.61 + 0.07 = 75.68 \text{ deg.}$$

$$\theta_5' = 106.60 + 0.08 = 106.97 \text{ deg.}$$

$$\theta_4' = 1.5(75.68) - .5(30) - 36.70 = 61.82 \text{ deg.}$$

These values are probably correct to about 0.01 deg., but we repeat the calculations once more to be sure. This time we find

$$\epsilon_1 = -0.000637 \qquad\qquad \epsilon_2 = -0.000159$$

$$a_1 = -7.1358 \qquad\qquad b_1 = 6.6952$$

$$a_2 = 2.1924 \qquad\qquad b_2 = 2.0431$$

$$\Delta\theta_3 = -0.0005 \text{ deg.} \qquad\qquad \Delta\theta_5 = 0.005 \text{ deg.}$$

Finally, we accept as correct to ± 0.01 deg. the values

$$\theta_3 = 75.68 \text{ deg.} \qquad\qquad \theta_5 = 106.97 \text{ deg.}$$

$$\theta_4 = 61.82 \text{ deg.}$$

(6) To continue the analysis, step θ_2 to a new value, θ_2 = 60 deg. and repeat (4) and (5). Use as starting estimates for θ_3 and θ_5 the values just calculated for the previous position.

$$\theta_3' = 75.68 \text{ deg.} \qquad\qquad \theta_5' = 106.97 \text{ deg.}$$

Use $\theta_4' = 1.5(75.68) - .5(60) - 36.70 = 46.82$ deg.

The computation routine suggested above is, of course, not one we would enjoy carrying out by hand very many times. If a full cycle analysis is required, the process should be programmed for automatic computation.

For this particular example, Fig. 2.7 shows the mechanism configuration at θ_2 = 270 deg.

2.7 Example Continued, Velocity and Acceleration Solutions via Kinematic Coefficients

(1) Differentiate equations (5), (6) and (7) of the previous article with respect to the input variable, θ_2

$$(-6\sin\theta_3)h_3 - (\sin\theta_4)h_4 + (7\sin\theta_5)h_5 = 0 \tag{1}$$

$$(6\cos\theta_3)h_3 + (\cos\theta_4)h_4 - (7\cos\theta_5)h_5 = 0 \tag{2}$$

$$h_4 = 1.5h_3 - 0.5 \tag{3}$$

where,

$$h_3 = d\theta_3/d\theta_2$$

$$h_4 = d\theta_4/d\theta_2$$

$$h_5 = d\theta_5/d\theta_2$$

(2) Angles θ_3, θ_4, θ_5 are known from the position solution. The above equations are linear in the unknowns h_3, h_4, h_5. Solve by any method preferred. One possibility is first to substitute in equations (1) and (2) the value of h_4 from equation (3). Equations (1) and (2) then become,

$$(-6\sin\theta_3 - 1.5\sin\theta_4)h_3 + (7\sin\theta_5)h_5 = -0.5\sin\theta_4 \tag{4}$$

$$(6\cos\theta_3 + 1.5\cos\theta_4)h_3 + (-7\cos\theta_5)h_5 = 0.5\cos\theta_4 \tag{5}$$

or

$$a_1 h_3 + b_1 h_5 = c_1 \tag{6}$$

$$a_2 h_3 + b_2 h_5 = c_2 \tag{7}$$

The coefficients of h_3 and h_5 are exactly the same as the coefficients of $\Delta\theta_3$ and $\Delta\theta_5$ in equations (14) and (15) of the position solution (previous article). We already have the values.

$a_1 = -7.1358$	$b_1 = 6.6952$
$a_2 = 2.1924$	$b_2 = 2.0431$

2.15

We evaluate

$$c_1 = -0.5\sin\theta_4 = -0.5\sin 61.82° = -0.44073$$

$$c_2 = 0.5\cos\theta_4 = 0.5\cos 61.82° = 0.23612$$

then complete the solution for h_3 and h_5.

$$h_3 = 0.08481 \qquad\qquad h_5 = 0.02456$$

From equation (3),

$$h_4 = 1.5(0.08481) - 0.5 = -0.37279$$

(3) Differentiate equations (1)-(3) with respect to θ_2.

$$(-6\sin\theta_3)h_3' - (6\cos\theta_3)h_3^2 - (\sin\theta_4)h_4' - (\cos\theta_4)h_4^2$$

$$+ (7\sin\theta_5)h_5' + (7\cos\theta_5)h_5^2 = 0 \qquad\qquad (8)$$

$$(6\cos\theta_3)h_3' - (6\sin\theta_3)h_3^2 + (\cos\theta_4)h_4' - (\sin\theta_4)h_4^2$$

$$- (7\cos\theta_5)h_5' + (7\sin\theta_5)h_5^2 = 0 \qquad\qquad (9)$$

$$h_4' = 1.5h_3' \qquad\qquad (10)$$

After substituting equation (10) into (8) and (9) and rearranging terms,

$$(-6\sin\theta_3 - 1.5\sin\theta_4)h_3' + (7\sin\theta_5)h_5'$$

$$= (6\cos\theta_3)h_3^2 + (\cos\theta_4)h_4^2 - (7\cos\theta_5)h_5^2 \qquad\qquad (11)$$

$$(6\cos\theta_3 + 1.5\cos\theta_4)h_3' + (-7\cos\theta_5)h_5'$$

$$= (6\sin\theta_3)h_3^2 + (\sin\theta_4)h_4^2 - (7\sin\theta_5)h_5^2 \qquad\qquad (12)$$

or,

$$a_1 h_3' + b_1 h_5' = d_1 \tag{13}$$

$$a_2 h_3' + b_2 h_5' = d_2 \tag{14}$$

Where the coefficients a_1, b_1, a_2, b_2 are the same as before.

$$d_1 = (6\cos 75.68°)(0.08481)^2 + (\cos 61.82°)(-0.37279)^2$$

$$- (7\cos 106.97°)(0.02456)^2$$

$$= 0.07754$$

$$d_2 = (6\sin 75.68°)(0.08481)^2 + (\sin 61.82°)(-0.37279)^2$$

$$- (7\sin 106.97°)(0.02456)^2$$

$$= 0.16028$$

The final results are

$$h_3' = 0.03126 \qquad h_5' = 0.04490 \qquad h_4' = 0.04689$$

(4) We can now compute the angular velocities and accelerations of 3, 4, and 5 for any given state of motion of input 2. For example, suppose $\dot{\theta}_2 = +20$ rad/sec and $\ddot{\theta}_2 = -300$ rad/sec^2.

$$\dot{\theta}_3 = h_3 \dot{\theta}_2 = (0.08481)(20) = 1.70 \text{ rad/sec}$$

$$\dot{\theta}_4 = h_4 \dot{\theta}_2 = (-0.37279)(20) = -7.46 \text{ rad/sec}$$

$$\dot{\theta}_5 = h_5 \dot{\theta}_2 = (0.02456)(20) = 0.49 \text{ rad/sec}$$

$$\ddot{\theta}_3 = h_3 \ddot{\theta}_2 + h_3' \dot{\theta}_2^2$$

$$= (0.08481)(-300) + (0.03126)(20)^2 = -12.9 \text{ rad/sec}^2$$

$$\ddot{\theta}_4 = h_5\ddot{\theta}_2 + h_4'\dot{\theta}_2^2$$

$$= (-0.37279)(-300) + (0.04689)(20)^2 = 130.6 \text{ rad/sec}^2$$

$$\ddot{\theta}_5 = h_5\ddot{\theta}_2 + h_5'\dot{\theta}_2^2$$

$$= (0.02456)(-300) + (0.04490)(20)^2 = 10.6 \text{ rad/sec}^2$$

2.8 Closed-Form Solutions

We have tended to emphasize an iterative approach to solving position equations because this is powerful and works well, even when two or more vector loops must be solved simultaneously. Closed-form solutions are always possible but may require more effort to develop than is worth expending. However, for any single-loop problem, a closed-form solution can be found with a reasonable effort. One example was the solution to the offset slider-crank problem in article 1.10, chapter 1.

For a second example consider the four-bar mechanism of Fig. 2.8. The vector loop equation is

$$\bar{r}_2 + \bar{r}_3 - \bar{r}_4 - \bar{r}_1 = 0 \tag{1}$$

The scalar position equations are

$$r_2\cos\theta_2 + r_3\cos\theta_3 - r_4\cos\theta_4 - r_1 = 0 \tag{2}$$

$$r_2\sin\theta_2 + r_3\sin\theta_3 - r_4\sin\theta_4 = 0 \tag{3}$$

With link 2 as the input (θ_2 known), the variable unknowns are θ_3 and θ_4. To eliminate one of these angles, say θ_3, do the following.

Rewrite equation (1)

$$\bar{r}_3 = \bar{r}_4 + \bar{r}_1 - \bar{r}_2 \tag{4}$$

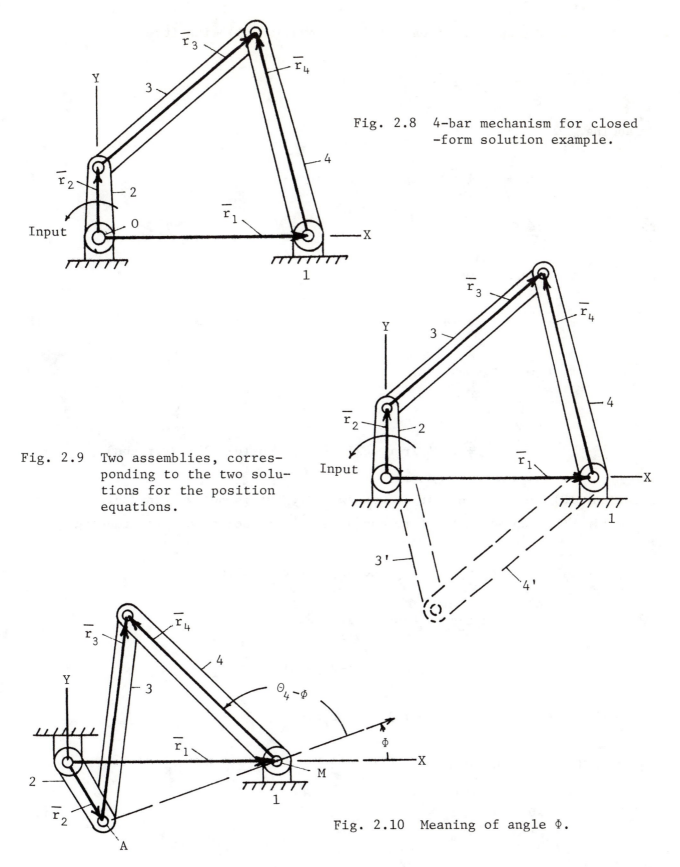

Fig. 2.8   4-bar mechanism for closed
-form solution example.

Fig. 2.9   Two assemblies, corresponding to the two solu-
tions for the position
equations.

Fig. 2.10   Meaning of angle Φ.

Form the dot product of each side of equation (1) with itself.

$$\bar{r}_3 \cdot \bar{r}_3 = (\bar{r}_4 + \bar{r}_1 - \bar{r}_2) \cdot (\bar{r}_4 + \bar{r}_1 - \bar{r}_2) \tag{5}$$

Expanding,

$$\bar{r}_3^2 = \bar{r}_4^2 + \bar{r}_1^2 + \bar{r}_2^2 + 2\bar{r}_4 \cdot \bar{r}_1 - 2\bar{r}_4 \cdot \bar{r}_2 - 2\bar{r}_1 \cdot \bar{r}_2 \tag{6}$$

But,

$$r_4 \cdot r_1 = r_4 r_1 \cos\theta_4 \tag{7}$$

$$r_4 \cdot r_2 = r_4 r_2 \cos(\theta_4 - \theta_2) \tag{8}$$

$$r_1 \cdot r_2 = r_1 r_2 \cos\theta_2 \tag{9}$$

So,

$$r_3^2 = r_4^2 + r_1^2 + r_2^2 + 2r_4 r_1 \cos\theta_4 - 2r_4 r_2 \cos(\theta_4 - \theta_2) - 2r_1 r_2 \cos\theta_2 \tag{10}$$

Equation (10) contains the single unknown, θ_4, but is not yet in explicit form. We continue by expanding the term $\cos(\theta_4 - \theta_2)$ and doing some rearranging.

$$r_3^2 = r_4^2 + r_1^2 + r_2^2 + 2r_4 r_1 \cos\theta_4 - 2r_4 r_2 \cos\theta_4 \cos\theta_2$$

$$- 2r_4 r_2 \sin\theta_4 \sin\theta_2 - 2r_1 r_2 \cos\theta_2 \tag{11}$$

$$(2r_4 r_1 - 2r_4 r_2 \cos\theta_2)\cos\theta_4 + (-2r_4 r_2 \sin\theta_2)\sin\theta_4$$

$$= (r_3^2 - r_4^2 - r_1^2 - r_2^2) + 2r_1 r_2 \cos\theta_2 \tag{12}$$

Equation (12) has the form

$$A \cos\theta_4 + B \sin\theta_4 = C \tag{13}$$

where,

$$A = 2r_4r_1 - 2r_4r_2\cos\theta_2 \tag{14}$$

$$B = -2r_4r_2\sin\theta_2 \tag{15}$$

$$C = (r_3^2 - r_4^2 - r_1^2 - r_2^2) + 2r_1r_2\cos\theta_2 \tag{16}$$

Write equation (13) as

$$\sqrt{A^2 + B^2}\left[\frac{A}{\sqrt{A^2 + B^2}}\cos\theta_4 + \frac{B}{\sqrt{A^2 + B^2}}\sin\theta_4\right] = C \tag{17}$$

Let,

$$A/\sqrt{A^2 + B^2} = \cos\phi \tag{18}$$

$$B/\sqrt{A^2 + B^2} = \sin\phi \tag{19}$$

Then we have

$$\sqrt{A^2 + B^2}(\cos\phi\cos\theta_4 + \sin\phi\sin\theta_4) = C$$

or,

$$\cos(\theta_4 - \phi) = C/\sqrt{A^2 + B^2} \tag{20}$$

2.9 Example, 4-Bar Closed-Form Solution (Fig. 2.8)

Dimensions: $r_1 = 3$, $r_2 = 1$, $r_3 = 3$, $r_4 = 3$ inches

Input: $\theta_2 = 90$ deg.

Using equations from article 2.8,

Equation (14): $A = 2r_4r_1 - 2r_4r_2\cos\theta_2$

$$= 2(3)(3) - 2(3)(1)\cos90° = 18$$

2.21

Equation (15): $B = -2r_4 r_2 \sin\theta_2$

$$= -2(3)(1)\sin 90° = -6$$

Equation (16): $C = r_3^2 - r_4^2 - r_1^2 - r_2^2 + 2r_1 r_2 \cos\theta_2$

$$= (3)^2 - (3)^2 - (3)^2 - (1)^2 + 2(3)(1)\cos 90° = -10$$

Equation (20): $\cos(\theta_4 - \phi) = C/\sqrt{A^2 + B^2}$

$$= -10/\sqrt{(18)^2 + (-6)^2} = -0.52705$$

Hence $(\theta_4 - \phi) = 121.81$ deg.   or   238.19 deg.

From equations (18) and (19)

$$\tan\phi = B/A = -6/18 = -.33333$$

So,

$$\phi = -18.43°$$

Hence,

$$\theta_4 = 121.81° - 18.43° = 103.38°$$

or

$$\theta_4 = 238.81° - 18.43° = 220.38°$$

Both of these solutions are real. They correspond to two different ways the mechanism can be assembled, as shown in Fig. 2.9. We take $\theta_4 = 103.38$ deg. to be the solution of interest in this problem.

To complete our problem we go back to the scalar position equations to find θ_3, as follows.

2.22

$$r_3 \cos\theta_3 = r_4 \cos\theta_4 + r_1 - r_2 \cos\theta_2$$

$$= (3)\cos 103.38° + 3 - (1)\cos 90° = 2.3058$$

$$r_3 \sin\theta_3 = r_4 \sin\theta_4 - r_2 \sin\theta_2$$

$$= (3)\sin 103.38° - (1)\sin 90° = 1.9186$$

$$\theta_3 = \text{arc } \tan(1.9186/2.3058) = \underline{39.76 \text{ deg.}}$$

2.10 Programming Closed-Form Solutions

The programming of closed-form solutions for automatic computation is very straightforward, once the steps have been worked out as in article 2.8. Notice that a closed-form solution does not mean deriving and writing a single explicit equation for each unknown. Rather, it means writing a series of explicit equations for calculations which eventually explicitly produce the values of the unknowns.

Every vector loop equation has two solutions. The programmer must be aware of this and be sure that the desired solution is the one selected. In the example of the previous article this decision should really be made at the point where $(\theta_4 - \phi)$ is calculated, instead of delaying until two values of θ_4 have been computed.

In order to make the choice between the two values of $(\theta_4 - \phi)$ we need to know what this angle means. To see this we go back to the definition of angle ϕ

$$\cos\phi = A/\sqrt{A^2 + B^2}$$

$$\sin\phi = B/\sqrt{A^2 + B^2}$$

$$B = -r_4 r_2 \sin\theta_2$$

$$A = r_4 r_1 - r_4 r_2 \cos\theta_2$$

Hence

$$\tan\phi = (-r_2\sin\theta_2)/(r_1 - r_2\cos\theta_2)$$

We see (Fig. 2.10) the meaning of ϕ, and hence of $(\theta_4 - \phi)$. The two assemblies of the 4-bar are distinguished from each other by the fact that $(\theta_4 - \phi)$ will be between $0°$ and $180°$ for the assembly shown solid and between $180°$ and $360°$ (or $0°$ and $-180°$) for the broken-line assembly. A computer or calculator via "arc cosine" will always give an angle between $0°$ and $180°$. Hence, for the 4-bar,

> (1) If we wish the solid line assembly we accept $(\theta_4 - \phi)$ as returned by the computer when we program
>
> $$(\theta_4 - \phi) = \text{arc cos } C/\sqrt{A^2 + B^2}$$
>
> (2) If we wish the broken line assembly we program
>
> $$(\theta_4 - \phi) = -\text{arc cos } C/\sqrt{A^2 + B^2}$$

2.11 Point Paths, Local Geometry

In chapter 1 there was a brief discussion of the kinematics of point motion. It was pointed out that, once the basic mechanism loops have been solved, the calculation of the position, velocity, and acceleration of any point belonging to one of the moving links is a straightforward process. Here we wish to point out that this _kinematic_ information enables us to make some statements about the geometry of the point path. Specifically, we can determine the slope of the path tangent and the location of the path center of curvature, via the following. (See Fig. 2.11)

> (1) The _velocity_ of the point is directed along the tangent to the path.
>
> (2) The _normal acceleration_ of the point is directed from the point toward the center of curvature of the path, and has the magnitude

2.24

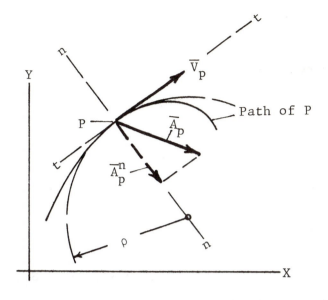

Fig. 2.11 Point path, velocity, acceleration.

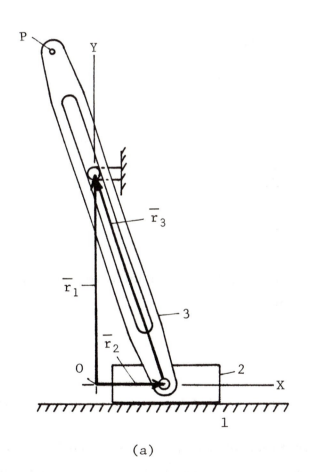

(a)

(b)

Fig. 2.12 Path curvature example.

2.25

There is a sign convention associated with equation (5) as follows.

ρ = + number means that the sense $\underline{\text{from the moving point}}$
$\underline{\text{to the path center}}$ of curvature is the same as $\overline{u}_n$.

ρ negative means $\overline{u}_n$ points away from the center of curvature.

The coordinates of the center of curvature are

$$X_c = X + \rho u_{nx} = X - \rho(f_y/f) \tag{6}$$

$$Y_c = Y + \rho u_{ny} = Y + \rho(f_x/f) \tag{7}$$

2.12 Example, Path Curvature (Fig. 2.12)

Locate the center of curvature of the path of point P for the mechanism position shown (r_2 = 1 in., r_1 = 3 in.)

(1) Basic mechanism vector loop, position solution.

$$\overline{r}_1 - \overline{r}_3 - \overline{r}_2 = 0$$

$$-r_3\cos\theta_3 - r_2 = 0$$

$$r_1 - r_3\sin\theta_3 = 0$$

$$-r_3\cos\theta_3 - 1 = 0$$

$$3 - r_3\sin\theta_3 = 0$$

Solution: θ_3 = 108.43 deg.

r_3 = 3.1623 in.

(2) Kinematic coefficients for the mechanism, using r_2 as input variable. Differentiate position equations with respect to r_2.

2.28

$$(r_3 \sin\theta_3)h_3 - (\cos\theta_3)f_3 = 1$$

$$(-r_3 \cos\theta_3)h_3 - (\sin\theta_3)f_3 = 0$$

where, $h_3 = d\theta_3/dr_2$, $f_3 = dr_3/dr_2$

$$3.000\ h_3 + 0.3161\ f_3 = 1$$

$$1.000\ h_3 - 0.9487\ f_3 = 0$$

Solution: $h_3 = 0.3000$

$$f_3 = 0.3162$$

(3) Differentiate again with respect to r_2.

$$(r_3 \sin\theta_3)h_3' - (\cos\theta_3)f_3' = (-r_3 \cos\theta_3)h_3^2 + 2f_3 h_3 \sin\theta_3 = C_1$$

$$(-r_3 \cos\theta_3)h_3' - (\sin\theta_3)f_3' = (-r_3 \sin\theta_3)h_3^2 + 2f_3 h_3 \cos\theta_3 = C_2$$

$$C_1 = 1.000(0.300)^2 - 2(0.3162)(0.3000)\sin 108.43 = -0.0900$$

$$C_2 = -3.000(.3000)^2 + 2(0.3162)(0.3000)\cos 108.43 = -0.3300$$

So, $3.000h_3' + 0.3161f_3' = -0.0900$

$$1.000h_3' - 0.9487f_3' = -0.3300$$

Solution: $h_3' = -0.0600$

$$f_3' = +0.2846$$

(4) Position of point P (Fig. 2.12(b)).

$$X_p = r_2 + r_5 \cos\theta_5$$

$$Y_p = r_5 \sin\theta_5$$

$$r_5 = 5\text{in.}, \quad r_2 = 1\text{in.}, \quad \theta_5 = \theta_3 = 108.45 \text{ deg.}$$

$$X_p = 1 + 5\cos108.43° = -0.5807\text{in.}$$

$$Y_p = 5\sin108.43° = +4.7436\text{in.}$$

(5) Kinematic coefficients associated with motion of P.

$$f_{xp} = dX_p/dr_2 = 1 - (r_5\sin\theta_3)h_3$$

$$= 1 - (5\sin108.43°)(0.3000) = -0.4231$$

$$f_{yp} = dY_p/dr_2 = (r_5\cos\theta_3)h_3$$

$$= (5\cos108.43°)(0.3000) = -0.4742$$

$$f'_{xp} = df_{xp}/dr_2 = (-r_5\sin\theta_3)h'_3 - (r_5\cos\theta_3)h_3^2$$

$$= (-5\sin108.43°)(-0.0600) - (5\cos108.43°)(0.3000)^2 = +0.4269 \text{ in.}^{-1}$$

$$f'_{yp} = (r_5\cos\theta_3)h'_3 - (r_5\sin\theta_3)h_3^2$$

$$= (5\cos108.43°)(-0.0600) - (5\sin108.43°)(0.3000)^2 = -0.3321 \text{ in.}^{-1}$$

(6) The unit vectors tangent and normal to the path of P.

$$\overline{u}_t = (f_{xp}/f_p)\overline{i} + (f_{yp}/f_p)\overline{j}$$

$$f_p = \sqrt{f_{xp}^2 + f_{yp}^2} = \sqrt{(-0.4231)^2 + (-0.4742)^2} = 0.6355$$

$$\overline{u}_t = (-0.4231/0.6355)\overline{i} + (-0.4742/0.6355)\overline{j}$$

$$= -0.6658\overline{i} - 0.7462\overline{j}$$

$$\overline{u}_n = 0.7462\overline{i} - 0.6658\overline{j}$$

(6) Path radius of curvature

$$\rho = f_p^{3}/(f'_{yp} f_{xp} - f'_{xp} f_{yp})$$

$$= (0.6355)^3/[(-0.3321)(-0.4231) - (0.4269)(-0.4742)]$$

$$= 0.748 \text{ in.}$$

Since this is a positive number, the unit normal vector points from P toward the path center of curvature.

(7) Coordinates for center of curvature, C.

$$X_c = X_p - \rho(f_{yp}/f_p)$$

$$= -0.5807 - 0.748(-0.4742/0.6355) = -0.023 \text{ in.}$$

$$Y_c = Y_p + \rho(f_{xp}/f_p)$$

$$= 4.7436 + 0.748(-0.4231/0.6355) = +4.246 \text{ in.}$$

Figure 2.12(b) shows a portion of the path plotted graphically as a rough check on these computations.

2.13 Two-Input Mechanisms

Devices with more than one degree of freedom can be usefully operated by appropriate independent inputs. For example, the 5-bar linkage of Fig. 2.13 might be operated by independently driving cranks 2 and 3 to produce a variety of motions of links 4 and 5. The kinematic analysis methods already discussed can be extended to such devices. For the 5-bar linkage shown in Fig. 2.13, the vector loop equation is

$$\bar{r}_2 + \bar{r}_4 - \bar{r}_5 - \bar{r}_3 - \bar{r}_1 = 0 \tag{1}$$

With the positions θ_3 and θ_2 of the inputs known, then the variable unknowns are θ_4 and θ_5. Nothing new would be involved in solving the position equations. Likewise the velocity and acceleration equations

$$\dot{\bar{r}}_2 + \dot{\bar{r}}_4 - \dot{\bar{r}}_5 - \dot{\bar{r}}_3 = 0 \tag{2}$$

$$\ddot{\bar{r}}_2 + \ddot{\bar{r}}_4 - \ddot{\bar{r}}_5 - \ddot{\bar{r}}_3 = 0 \tag{3}$$

each contain two variable unknowns and can be solved just as in any of our previous examples.

The only thing that would need modifying for multi-degree-of-freedom systems would be our concept of <u>kinematic</u> <u>coefficients</u>. For our 5-bar example, in the position equations, the dependent variables, θ_4 and θ_5, are functions of the two <u>independent</u> <u>absolute</u> <u>variables</u>, θ_2 and θ_3.

Hence,

$$d\theta_4/dt = (\partial\theta_4/\partial\theta_2)(d\theta_2/dt) + (\partial\theta_4/\partial\theta_3)(d\theta_3/dt) \tag{4}$$

or,

$$\dot{\theta}_4 = h_{42}\dot{\theta}_2 + h_{43}\dot{\theta}_3 \tag{5}$$

where

$$h_{42} = \partial\theta_4/\partial\theta_2 \quad \text{and} \quad h_{43} = \partial\theta_4/\partial\theta_2$$

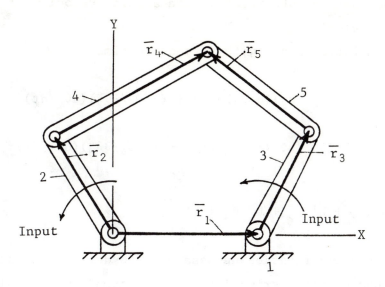

Fig. 2.13 A two-input (two-degree-of-freedom) mechanism.

are two kinematic coefficients relating the angular velocity of link 4 to the two input angular velocities

Upon differentiating again with respect to time,

$$d^2\theta_4/dt^2 = (\partial\theta_4/\partial\theta_2)(d^2\theta_2/dt^2) + (\partial^2\theta_4/\partial\theta_2^2)(d\theta_2/dt)^2$$

$$+ (\partial\theta_4/\partial\theta_3)(d^2\theta_3/dt^2) + (\partial^2\theta_4/\partial\theta_3^2)(d\theta_3/dt)^2 \tag{6}$$

or,

$$\ddot{\theta}_4 = h_{42}\ddot{\theta}_2 + h'_{42}\dot{\theta}_2^2 + h_{43}\ddot{\theta}_3 + h'_{43}\dot{\theta}_3^2 \tag{7}$$

We see that the only modification necessary is to define kinematic coefficients as first and second _partial_ derivatives of the dependent variables with respect to the input variables.

To generate the kinematic coefficients for the 5-bar, two-input mechanism we would proceed as follows.

(1)   Write the scalar position equations, and solve for θ_4 and θ_5.

$$r_2\cos\theta_2 + r_4\cos\theta_4 - r_5\cos\theta_5 - r_3\cos\theta_3 - r_1 = 0 \tag{8}$$

$$r_2\sin\theta_2 + r_4\sin\theta_4 - r_5\sin\theta_5 - r_3\sin\theta_3 = 0 \tag{9}$$

(2)   Differentiate partially with respect to θ_2.

$$-r_2\sin\theta_2 - (r_4\sin\theta_4)h_{42} + (r_5\sin\theta_5)h_{52} = 0 \tag{10}$$

$$r_2\cos\theta_2 + (r_4\cos\theta_4)h_{42} - (r_5\cos\theta_5)h_{52} = 0 \tag{11}$$

Solve for h_{42} and h_{52}.

(3)   Differentiate partially with respect to θ_3.

$$(-r_4\sin\theta_4)h_{43} + (r_5\sin\theta_5)h_{53} + r_3\sin\theta_3 = 0 \tag{12}$$

$$(r_4\cos\theta_4)h_{43} - (r_5\cos\theta_5)h_{53} - r_3\cos\theta_3 = 0 \tag{13}$$

Solve for h_{43} and h_{53}.

(4) Differentiate equations (10) and (11) partially with respect to θ_2.

$$-r_2\cos\theta_2 - (r_4\sin\theta_4)h'_{42} - (r_4\cos\theta_4)h^2_{42} + (r_5\sin\theta_5)h'_{52}$$

$$+ (r_5\cos\theta_5)h^2_{52} = 0 \tag{14}$$

$$-r_2\sin\theta_2 + (r_4\cos\theta_4)h'_{42} - (r_4\sin\theta_4)h^2_{42} - (r_5\cos\theta_5)h_{52}$$

$$+ (r_5\sin\theta_5)h^2_{52} = 0 \tag{15}$$

Solve for h'_{42} and h'_{52}.

(5) Differentiate equations (80) and (81) partially with respect to θ_3.

$$(-r_4\sin\theta_4)h'_{43} - (r_4\cos\theta_4)h^2_{43} + (r_5\sin\theta_5)h'_{53} + (r_5\cos\theta_5)h^2_{53}$$

$$+ r_3\cos\theta_3 = 0 \tag{16}$$

$$(r_4\cos\theta_4)h'_{43} - (r_4\sin\theta_4)h^2_{43} - (r_5\cos\theta_5)h'_{53} + (r_5\sin\theta_5)h^2_{53}$$

$$+ r_3\sin\theta_3 = 0 \tag{17}$$

In the interest of computation efficiency it should be observed that the above equations to be solved for kinematic coefficients are of the form

$$a_1 h_{42} + b_1 h_{52} = c_{12}$$

$$a_2 h_{42} + b_2 h_{52} = c_{22}$$

$$a_1 h_{43} + b_1 h_{43} = c_{13}$$

$$a_2 h_{43} + b_2 h_{43} = c_{23}$$

$$a_1 h'_{42} + b_1 h'_{52} = d_{12}$$

$$a_2 h'_{42} + b_2 h'_{52} = d_{22}$$

$$a_1 h'_{43} + b_1 h'_{53} = d_{13}$$

$$a_2 h'_{43} + b_2 h'_{53} = d_{23}$$

or. letting $[A] = \begin{bmatrix} a_1 & b_1 \\ a_2 & b_2 \end{bmatrix}$

$$[A] \begin{bmatrix} h_{42} & h_{43} \\ h_{52} & h_{53} \end{bmatrix} = \begin{bmatrix} c_{12} & c_{13} \\ c_{22} & c_{23} \end{bmatrix}$$

$$[A] \begin{bmatrix} h'_{42} & h'_{43} \\ h'_{52} & h'_{53} \end{bmatrix} = \begin{bmatrix} d_{12} & d_{13} \\ d_{22} & d_{23} \end{bmatrix}$$

2.14 Checking Results

The results of a computerized kinematic analysis are ordinarily presented as columns of numbers or as plotted graphs. In either case the results should be examined carefully for assurance that no errors have been made. Never blindly accept such results without checking. Two types of checks are strongly recommended.

Output consistency checks

Examine the output information to see whether it is consistent. For example, the table of Fig. 2.14 shows the printed output for a mechanism in which θ_2 is the input variable, θ_3 and r_3 are the position variables computed, and h_3, h_3', f_3 and f_3' are the corresponding kinematic coefficients computed. We can very quickly check things like the following, without any calculations.

    (1) With increasing θ_2 is h_3 positive while θ_3 is increasing and negative while θ_3 is decreasing?

    (2) Is h_3' positive while h_3 is increasing?

    (3) Is h_3' zero where h_3 is maximum or minimum?

If the answer to any of these questions is "no," then something is wrong and we should look for the trouble.

We can also do some simple arithmetic to check on the consistency of the magnitudes of the numbers.

    (1) Pick a short range of θ_2, say from $\theta_2 = 70°$ to $90°$ and calculate $\Delta\theta_3/\Delta\theta_2 = 6.55°/20° = 0.327$. This should be a rather close approximation to the computed value of h_3 at $\theta_2 = 80°$ (the midpoint of the interval tested).

    (2) For the same interval calculate $\Delta h_3/\Delta\theta_2$. This should approximate the value of h_3' computed at the midpoint of the interval.

Independent checks

Normally, if the consistency checks suggested above look good, we can be fairly confident our program is correct. However, to increase our confidence greatly, we should do a little checking by some independent method. It is suggested that a scale layout of the mechanism be made for one position of the input link. The position variables can be measured from the drawing and compared with the computed results. If we then, on the same drawing, locate the instant centers of the mechanism, we can use these to calculate the first-order kinematic coefficients for comparison with the computer results. See Fig. 2.15 for an example.

θ_2	θ_3	r_3	h_3	f_3	h_3'	f_3'
degrees	degrees	inches		inches		inches
0.000	72.008	6.633	.140	2.714	.313	−1.110
10.000	73.656	7.089	.188	2.501	.242	−1.323
20.000	75.728	7.504	.225	2.254	.188	−1.497
30.000	78.131	7.874	.254	1.980	.148	−1.641
40.000	80.793	8.194	.277	1.683	.116	−1.758
50.000	83.659	8.460	.295	1.368	.091	−1.851
60.000	86.683	8.670	.309	1.038	.070	−1.922
70.000	89.831	8.822	.320	.698	.053	−1.972
80.000	93.072	8.914	.328	.351	.038	−2.002
90.000	96.379	8.944	.333	−.000	.025	−2.012
100.000	99.731	8.914	.337	−.351	.012	−2.002
110.000	103.103	8.822	.338	−.698	−.000	−1.972
120.000	106.475	8.670	.336	−1.038	−.013	−1.922
130.000	109.823	8.460	.333	−1.368	−.028	−1.851
140.000	113.123	8.194	.327	−1.683	−.044	−1.758
150.000	116.344	7.874	.317	−1.980	−.064	−1.641
160.000	119.453	7.504	.304	−2.254	−.090	−1.497
170.000	122.403	7.089	.285	−2.501	−.123	−1.323
180.000	125.138	6.633	.260	−2.714	−.167	−1.110
190.000	127.579	6.144	.226	−2.885	−.229	−.846
200.000	129.617	5.629	.179	−3.005	−.317	−.510
210.000	131.096	5.099	.113	−3.057	−.444	−.068
220.000	131.792	4.567	.021	−3.019	−.629	.538
230.000	131.377	4.052	−.111	−2.855	−.891	1.391
240.000	129.397	3.581	−.295	−2.513	−1.222	2.589
250.000	125.287	3.189	−.536	−1.930	−1.512	4.136
260.000	118.590	2.924	−.802	−1.069	−1.447	5.673
270.000	109.471	2.828	−1.000	.000	−.707	6.364
280.000	99.177	2.924	−1.026	1.069	.405	5.672
290.000	89.531	3.189	−.881	1.930	1.152	4.136
300.000	81.809	3.581	−.659	2.513	1.321	2.590
310.000	76.346	4.052	−.439	2.855	1.167	1.391
320.000	72.909	4.567	−.256	3.019	.929	.538
330.000	71.096	5.099	−.113	3.057	.711	−.068
340.000	70.530	5.629	−.005	3.005	.538	−.510
350.000	70.910	6.144	.077	2.885	.409	−.846
360.000	72.008	6.633	.140	2.714	.313	−1.110

Fig. 2.14 Computed results of a kinematic analysis.

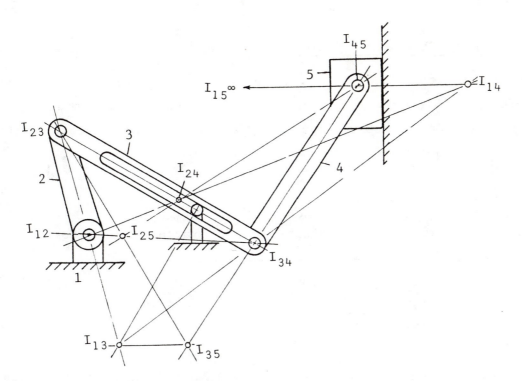

(a) Mechanism and instant center locations.

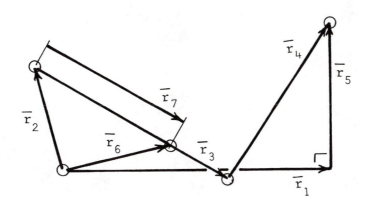

(b) Vector loops.

Fig. 2.15 Illustrating the determination of first-order kinematic coefficients by instant centers.

$$h_3 = \omega_3/\omega_2 = (I_{12}I_{23})/(I_{13}I_{23}) = 0.48$$

$$h_4 = \omega_4/\omega_2 = (I_{12}I_{24})/(I_{14}I_{24}) = -0.30$$

$$f_5 = V_5/\omega_2 = (I_{12}I_{25}) = 0.34 \text{ in.}$$

2.15 On Estimating Position Variables

Earlier, in article 1.8, it was suggested that when we do the iterative solution for position variables, we make our first estimates from a scale drawing. Then, for each successive input position, we let the values from the previous position be the starting estimates.

Hypothetical Example

Position	Input Variable θ_2	Estimated Variable θ_3'	Computed Variable θ_3	Coefficient h_3
1	$0°$	$30°$	$32.67°$	0.59
2	$10°$	$32.67°$	$38.13°$	
3	$20°$	$38.13°$		

This process can be improved in a very simple way, using the first-order kinematic coefficients. In the hypothetical example, above, at the end of step 1, create the next estimate for θ_3 as follows.

$$\text{New est } \theta_3 = \text{current } \theta_3 + h_3 \Delta\theta_2$$

$$= 32.67° + 0.59(10°) = 38.57°$$

This is a much closer estimate of the correct value of θ_3 in position 2, because it takes into account approximately how fast link 3 is turning between positions 1 and 2.

Not only does this improved estimate cut down somewhat on the iterations needed, it also cures an occasional computational problem which arises in certain special cases where the mechanism "folds." For an example, consider the crossed parallelogram 4-bar of Fig. 2.16. Suppose that we started the analysis at $\theta_2 = 5°$, to avoid starting in the folded configuration. If we use steps of $\Delta\theta_2 = 10°$ everything should go all right until we reach $\theta_2 = 185°$. If, for this position, we used as an estimate of θ_4 the previous value (for $\theta_2 = 175°$), we would find the solution converging to the open (parallelogram) configuration of the linkage. If we use the suggested improved estimate of θ_4 the solution will

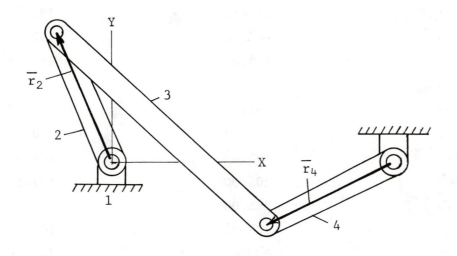

Fig. 2.16 Crossed parallelogram 4-bar.

continue to converge to the crossed configuration. These results are shown in
the following table.

Θ_2, deg.	Θ_4, deg. Based on simple estimate	Θ_4, deg. Based on improved estimate
105	208.69	208.70
115	203.98	203.98
125	199.69	199.69
135	195.72	195.72
145	192.00	192.00
155	188.45	188.45
165	185.03	185.03
175	181.67	181.67
185	185.00	178.33
195	195.00	174.97
205	205.00	171.55
215	215.00	168.00
225	225.00	164.28
235	235.00	160.31

E2.1 For the mechanism shown define suitable vectors and write vector loop
      equations for a kinematic analysis. Carefully note the variables involved.

      Assuming the rotation of link 2 to be the input, outline the procedure
      for the kinematic analysis. Make a flow diagram for the computation.

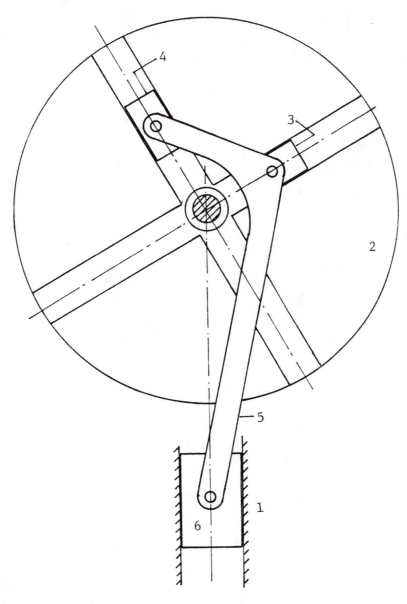

2.43

E2.2 For the mechanism shown define suitable vectors and write all equations needed for a kinematic analysis. Carefully note the variables involved.

Assuming the rotation of link 2 to be the input variable, outline the procedure for the kinematic analysis. Make a flow diagram for the computation.

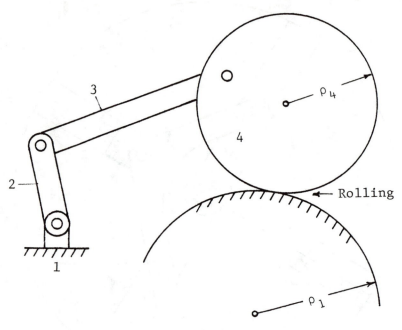

E2.3 In the mechanism shown the three gears are of equal size.

(a) Choose a coordinate system, define vectors, and write the equations needed for a complete kinematic analysis.

(b) Program the computations and explore the effect of varying the distance between the fixed pivots.

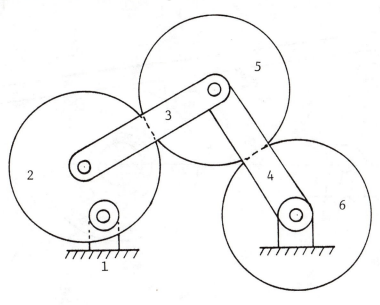

E2.4 (a) Define suitable vectors and write the equations necessary for a kinematic analysis of the mechanism shown.

(b) Assume the length of crank 2 to be 3 in., the distance between fixed
pivots to be 8 in., and the radius of gear 5 to be 2 in. Write and
run a computer program for the kinematic analysis. Assume crank 2 to
be the input.

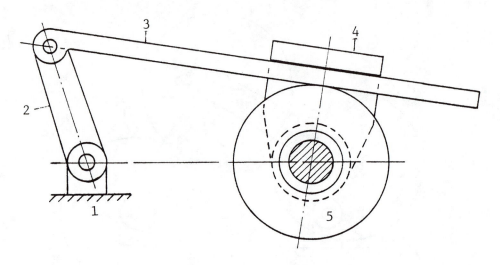

E2.5 (a) Define suitable vectors and write equations for a kinematic analysis
         of the mechanism shown. Assume crank 2 to be the input.

      (b) Write equations needed to compute the location of point B.

      (c) Write equations needed to compute the location of the center of cur-
          vature for the path of B.

      (d) For Θ_2 = 210 deg. execute the above computations.

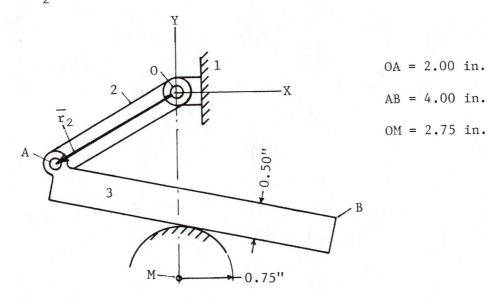

OA = 2.00 in.

AB = 4.00 in.

OM = 2.75 in.

E2.6 For the mechanism shown define suitable vectors and write all equations
      needed for a kinematic analysis. Carefully note the variables involved.

      Assuming the rotation of link 2 to be the input variable, outline the
      procedure for the kinematic analysis.

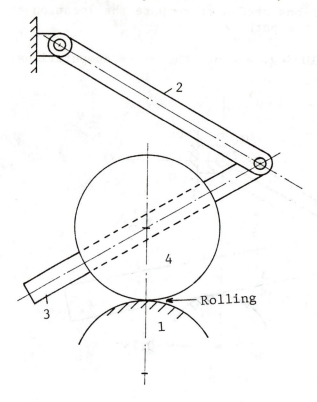

E2.7 Consider link 2 to be the input. Define suitable vectors and write vector equations for a kinematic anlaysis. Include the determination of the
relative velocity between 3 and 2 at points D and E.

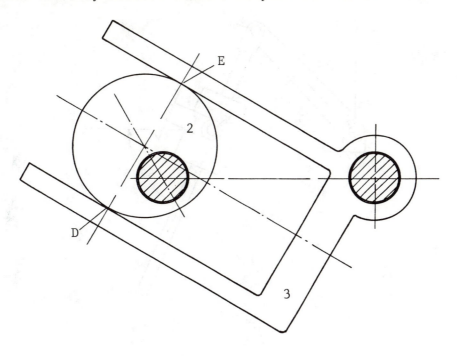

E 2.8 The mechanism is designed such that $\omega_3 = \omega_2/2$ continuously. AB = 6 in.
        OA = OM = 3 in. For the position indicated, compute the location of
        the center of curvature for the path of point B.

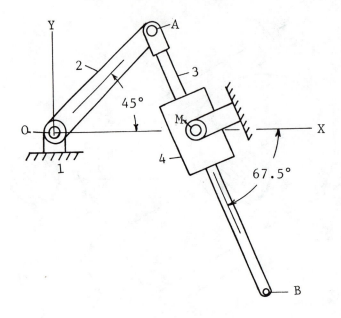

CHAPTER 3

FORCE ANALYSIS I

3.1 Introduction

For a given mechanism in a known state of motion the determination of
forces exerted by different links on each other can be done by a straightforward
application of Newton's laws. By a given mechanism we mean one for which all
pertinent constants (dimensions, masses, center-of-mass locations, moments of
inertia) are known. By a known state of motion we mean a known position,
velocity and acceleration of the input link.

A necessary preliminary to the force analysis is the kinematic analysis of
the mechanism. Specifically we need to determine the angular accelerations of
all moving parts and the accelerations of all mass centers.

3.2 Newton's Laws (Fig. 3.1)

For a body in plane motion

$$\Sigma \overline{F}_e = m\overline{A}_g \qquad (1)$$

or

$$\Sigma F_{ex} = mA_{gx} \qquad (2)$$

$$\Sigma F_{ey} = mA_{gy} \qquad (3)$$

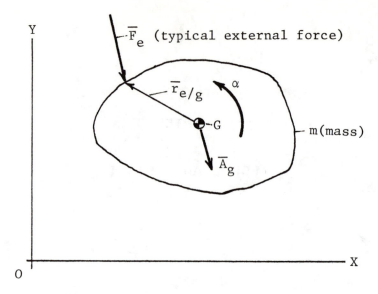

Fig. 3.1 Body in plane motion.

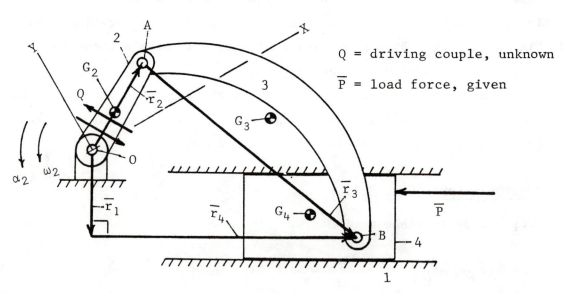

Q = driving couple, unknown

$\overline{P}$ = load force, given

Fig. 3.2 Offset slider-crank, to illustrate force analysis approach.

and

$$\dot{\Sigma M}_g = I\alpha \tag{4}$$

where,

$\overline{F}_e$ = a typical <u>external</u> force.

$\Sigma \overline{F}_e$ = summation of <u>all</u> external forces.

F_{ex}, F_{ey} = X,Y components of $\overline{F}_e$.

m = mass of the body.

$\overline{A}_g$ = acceleration of the center of mass.

A_{gx}, A_{gy} = X,Y components of $\overline{A}_g$.

ΣM_g = summation of moments of all external forces, around axis through center of mass perpendicular to the plane of motion (the X,Y plane).

I = moment of inertia of the body around the axis <u>through the center of mass</u> and perpendicular to the plane of motion.

α = angular acceleration of the body.

A complete force analysis for a machine would consist of writing the three scalar equations (equations (2), (3), (4)) for each moving part of the machine, then solving the resulting set of equations for the unknowns. For computer solutions (non-graphical solutions) we are suggesting that all vectors in the plane be described by X,Y components. This means, for example, that M_g in equation (4) would be set up as follows,

$$\overline{M}_g = \overline{r}_{e/g} \times \overline{F}_e = \left[(r_{e/g}\cos\theta_e)\overline{i} + (r_{e/g}\sin\theta_e)\overline{j} \right] \times \left[F_{ex}\overline{i} + F_{ey}\overline{j} \right]$$

$$= \left[(r_{e/g}\cos\theta_e)F_{ey} - (r_{e/g}\sin\theta_e)F_{ex} \right]\overline{k} \tag{5}$$

where $\bar{i}$, $\bar{j}$ and $\bar{k}$ are unit vectors in the X,Y,Z directions. Then,

$$\Sigma M_g = \Sigma \left[(r_{e/g} \cos\theta_e) F_{ey} - (r_{e/g} \sin\theta_e) F_{ex} \right] \tag{6}$$

3.3 Example. Setting Up Equations (Fig. 3.2)

Assume that the offset slider-crank mechanism shown in Fig. 3.2 is being driven by a known force, $\bar{P}$, applied to the slider as shown. An unknown torque, Q, is applied to crank 2 by some external agent. Further assume that all physical constants (lengths, masses, moments of inertia, locations for centers of mass) are known and that a kinematic loop analysis for the mechanism has been completed. Friction and weight forces are to be neglected in this example. The problem is to determine Q and all the reaction forces between links.

(1) Make free-body sketches of the moving links (Fig. 3.3). This is a very important step, so will be outlined in some detail as follows.

(a) Make a freehand picture of each link.

(b) Starting with any one of the link pictures, identify the external forces and couples. Assign appropriate symbolic "names" to each force or couple. In this example we started with link 2. We recognized the existence of couple Q, the force $\bar{F}_{12}$ exerted <u>by</u> link 1 <u>on</u> link 2, and the force $\bar{F}_{32}$ exerted <u>by</u> link 3 <u>on</u> link 2. Notice that we make no attempt to guess at the lines of action of $\bar{F}_{12}$ or $\bar{F}_{32}$, nor to guess whether X and Y components might be positive or negative. We do recognize that $\bar{F}_{12}$ acts through the pin center, "O," and $\bar{F}_{32}$ acts through the pin center "B." We simply sketch broken vectors to indicate that both magnitudes and directions are unknown.

(c) Move to another link picture and do the same thing. In this example we went to link 3 next. We observe that links 2 and 4 each exert forces on 3 through the connecting pin joints. We recognize that $\bar{F}_{23} = -\bar{F}_{32}$, so instead of choosing a new "name" for the force exerted <u>by</u> 2 <u>on</u> 3 we label it "$-\bar{F}_{32}$."

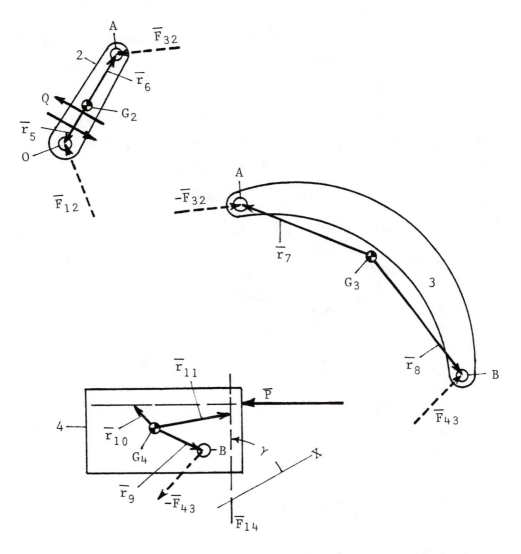

Fig. 3.3 Freebody sketches.

(d) Moving to link 4, we recognize that, in addition to the given force $\overline{P}$, we have forces $\overline{F}_{34}$ and $\overline{F}_{14}$. Again, since we have already symbolized $\overline{F}_{43}$ we label the force of 3 on 4 "$-\overline{F}_{43}$." The force $\overline{F}_{14}$ is known to be normal to the contacting surfaces, but its location may not be immediately known. We use a straight, broken line to indicate the known line of action <u>direction</u> and unknown location.

(e) Now identify and assign names to all vectors <u>from</u> the center of mass of each link <u>to</u> points on the line of action of each external force on that link. (In Fig. 3.3, vectors $\overline{r}_5$ through $\overline{r}_{11}$.)

The free-body sketches are now complete.

(2) Write the force equations (equations (2) and (3)) for each link.

(a) Link 2

$$F_{12x} + F_{32x} = m_2 A_{g2x} \tag{1}$$

$$F_{12y} + F_{32y} = m_2 A_{g2y} \tag{2}$$

(b) Link 3

$$-F_{32x} + F_{43x} = m_3 A_{g3x} \tag{3}$$

$$-F_{32y} + F_{43y} = m_3 A_{g3y} \tag{4}$$

(c) Link 4

$$P_x - F_{43x} + F_{14x} = m_4 A_{g4x} \tag{5}$$

$$P_y - F_{43y} + F_{14y} = m_3 A_{g3y} \tag{6}$$

(3) Write the moment equation for each link.

(a) Link 2

$$Q + (r_5\cos\theta_5)F_{12y} - (r_5\sin\theta_5)F_{12x} + (r_6\cos\theta_6)F_{32y}$$

$$- (r_6\sin\theta_6)F_{32x} = I_2\alpha_2 \qquad (7)$$

(b) Link 3

$$(r_7\cos\theta_7)(-F_{32y}) - (r_7\sin\theta_7)(-F_{32x}) + (r_8\cos\theta_8)F_{43y}$$

$$- (r_8\sin\theta_8)F_{43x} = I_3\alpha_3 \qquad (8)$$

(c) Link 4

$$(r_9\cos\theta_9)(-F_{43y}) - (r_9\sin\theta_9)(-F_{43x}) + (r_{11}\cos\theta_{11})F_{14y}$$

$$- (r_{11}\sin\theta_{11})F_{14x} + (r_{10}\cos\theta_{10})P_y$$

$$- (r_{10}\sin\theta_{10})P_x = I_4\alpha_4 \qquad (9)$$

(4) The known scalar quantities in the force and moment equations would be

(a) Masses and moments of inertia m_2, m_3, m_4, I_2, I_3, I_4.

(b) All vector lengths, r_5 through r_{10}.

(c) All angles θ_5 through θ_{11}.

(d) The given force, P.

(5) The unknown scalar quantities are

$$Q, F_{12x}, F_{12y}, F_{32x}, F_{32y}, F_{43x}, F_{43y}, F_{14x}, F_{14y} \text{ and } r_{11}.$$

So, we appear to have written <u>nine equations</u> in <u>ten unknowns</u>. However, F_{14x} and F_{14y} are not independent unknowns since we know the direction of $\overline{F}_{14}$.

$$F_{14x} = F_{14}\cos\gamma, \quad \text{and} \quad F_{14y} = F_{14}\sin\gamma$$

If we make this substitution in equations (6) and (9), then we have only nine unknowns in our nine equations.

$$Q, \; F_{12x}, \; F_{12y}, \; F_{32x}, \; F_{32y}, \; F_{43x}, \; F_{43y}, \; F_{14} \; \text{and} \; r_{11}$$

We should also notice that this set of equations is _nearly_ _linear_ in the nine listed unknowns. Equation (9) is non-linear because two of the unknowns (F_{14} and r_{11}) appear as a product. However, if we let the product ($r_{11}F_{14}$) replace r_{11} in our list of unknowns, we will have a completely linear set of nine equations in nine unknowns.

$$Q, \; F_{12x}, \; F_{12y}, \; F_{32x}, \; F_{32y}, \; F_{43x}, \; F_{43y}, \; F_{14} \; \text{and} \; (r_{11}F_{14})$$

Upon arranging these equations in "standard" linear form (known terms on the right, unknowns written in the same order in all equations) we have

$$F_{12x} + F_{32x} = m_2 A_{g2x} \tag{10}$$

$$F_{12y} + F_{32y} = m_2 A_{g2y} \tag{11}$$

$$-F_{32x} + F_{43x} = m_3 A_{g3x} \tag{12}$$

$$-F_{32y} + F_{43y} = m_3 A_{g3y} \tag{13}$$

$$-F_{43x} + (\cos\gamma)F_{14} = m_4 A_{g4x} - P_x \tag{14}$$

$$-F_{43y} + (\sin\gamma)F_{14} = m_4 A_{g4y} - P_y \tag{15}$$

$$Q - (r_5\sin\theta_5)F_{12x} - (r_6\sin\theta_6)F_{32x}$$

$$+ (r_5\cos\theta_5)F_{12y} + (r_6\cos\theta_6)F_{32y} = I_2\alpha_2 \tag{16}$$

$$(r_7 \sin\theta_7)F_{32x} - (r_7 \cos\theta_7)F_{32y}$$

$$- (r_8 \sin\theta_8)F_{43x} + (r_8 \cos\theta_8)F_{43y} = I_3\alpha_3 \tag{17}$$

$$(r_9 \sin\theta_9)F_{43x} - (r_9 \cos\theta_9)F_{43y}$$

$$+ (-\sin\theta_{11}\cos\gamma + \cos\theta_{11}\sin\gamma)r_{11}F_{14} = I_4\alpha_4 - (r_{10}\cos\theta_{10})P_y$$

$$+ (r_{10}\sin\theta_{10})P_x \tag{18}$$

Equations (10)-(18) represent one formulation of the force analysis problem. A slight variation of this, which may be advantageous at times, will be pointed out in the next article.

3.4 The Moment Equations

In the previous article the moment equations were written by taking moments around the axis through the center of mass (G) of each link. Some simplification may be introduced by taking moments around a different axis. See Fig. 3.4.

$$\Sigma\overline{M}_a = \Sigma(\overline{r}_{e/a} \times \overline{F}_e)$$

$$= \Sigma\left[(\overline{r}_{g/a} + \overline{r}_{e/g}) \times \overline{F}_e\right]$$

$$= \Sigma(\overline{r}_{g/a} \times \overline{F}_e) + \Sigma(\overline{r}_{e/g} \times \overline{F}_e)$$

$$= \overline{r}_{g/a} \times \Sigma\overline{F}_e + \Sigma(\overline{r}_{e/g} \times \overline{F}_e)$$

But,

$$\Sigma\overline{F}_e = m\overline{A}_g$$

and

$$\Sigma(\overline{r}_{e/g} \times \overline{F}_e) = \overline{I\alpha}$$

Hence,

$$\Sigma\overline{M}_a = \overline{r}_{g/a} \times m\overline{A}_g + \overline{I\alpha} \tag{1}$$

3.9

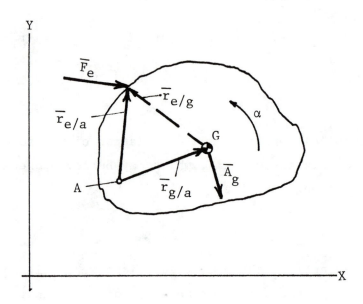

Fig. 3.4   Moments around axis not through the center of mass.

In the illustrative example of article 3.3 we might have chosen to write moment equations as follows (see free-body sketches in Fig. 3.5).

Link 2, moments around O:

$$Q - (r_2\sin\theta_2)F_{32x} + (r_2\cos\theta_2)F_{32y} = -(r_5\sin\theta_5)m_2A_{g2x}$$

$$+ (r_5\cos\theta_5)m_2A_{g2y} + I_2\alpha_2 \qquad (2)$$

Link 3, moments around A:

$$-(r_3\sin\theta_3)F_{43x} + (r_3\cos\theta_3)F_{43y} = -(r_6\sin\theta_6)m_3A_{g3x}$$

$$+ (r_6\cos\theta_6)m_3A_{g3y} + I_3\alpha_3 \qquad (3)$$

Link 4, moments around B:
Defining $\overline{r}_8$ and $\overline{r}_9$ to be perpendicular to $\overline{P}$ and $\overline{F}_{14}$, respectively, and taking note of the obvious fact that $\alpha_4 = 0$, we have

$$r_9F_{14} = -(r_7\sin\theta_7)m_4A_{g4x} + (r_7\cos\theta_7)m_4A_{g4y} - r_8P \qquad (4)$$

3.5 Numerical Example. Inverted Slider-Crank (Fig. 3.6)

The inverted slider-crank mechanism described in Fig. 3.6 is to be analyzed to determine the driving couple, Q, and the forces at the various joints. The crank is rotating at a constant speed and there is no external loading of the mechanism.

(A) Do the kinematic analysis of the basic mechanism loop (Fig. 3.7).

(1) $\overline{r}_2 - \overline{r}_3 - \overline{r}_1 = 0$

$r_2\cos\theta_2 - r_3\cos\theta_3 = 0$

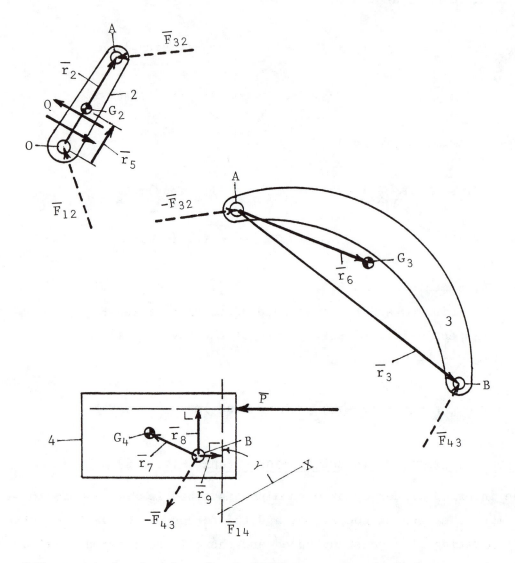

Fig. 3.5 Freebody sketches for a different set of moment equations.

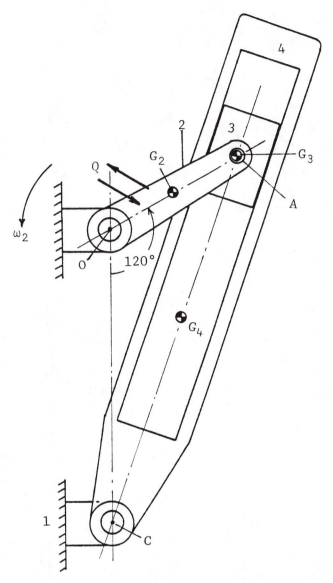

Fig. 3.6 Inverted slider-crank mechanism, for force analysis example.

OA = 6 in., OC = 12 in., OG_2 = 3 in., CG_4 = 9 in.

m_2 = 3 lb_m, m_3 = 2 lb_m, m_4 = 8 lb_m, I_2 = 12 $lb_m in^2$,

I_3 = 0.5 $lb_m in^2$, I_4 = 200 $lb_m in^2$,

ω_2 = 40 rad/sec, constant.   Q = driving couple.

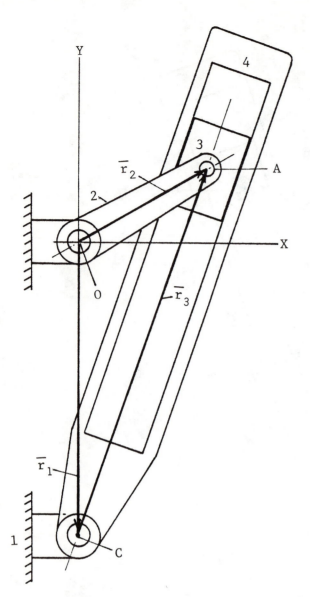

Fig. 3.7 Basic vector loop for kinematic analysis.

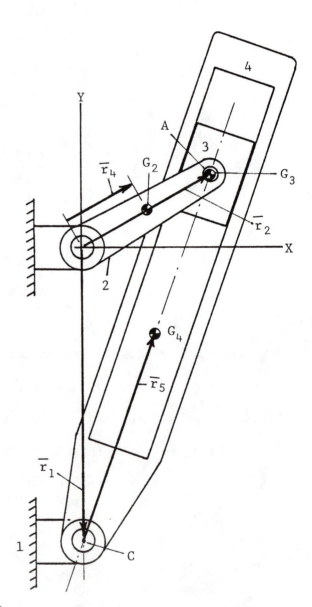

Fig. 3.8 Vectors defining locations
of mass centers.

$$r_2 \sin\theta_2 - r_3 \sin\theta_3 + r_1 = 0$$

$$(r_2 = 6 \text{ in.}, \quad r_1 = 12 \text{ in.}, \quad \theta_2 = 30 \text{ deg.})$$

Solution: $r_3 = 15.875$ in.

$$\theta_3 = 70.894 \text{ deg.}$$

(2) $\quad (r_3 \sin\theta_3) h_3 + (-\cos\theta_3) f_3 = r_2 \sin\theta_2$

$$(-r_3 \cos\theta_3) h_3 + (-\sin\theta_3) f_3 = -r_2 \cos\theta_2$$

where $\quad h_3 = d\theta_3/d\theta_2, \quad f_3 = dr_3/d\theta_2$

Solution: $h_3 = 0.28570$

$$f_3 = 3.92784 \text{ in.}$$

(3) $\quad (r_3 \sin\theta_3) h_3' + (-\cos\theta_3) f_3' = -r_3 \cos\theta_3 h_3^2 - 2f_3 h_3 \sin\theta_3 + r_2 \cos\theta_2$

$$(-r_3 \cos\theta_3) h_3' + (-\sin\theta_3) f_3' = -r_3 \sin\theta_3 h_3^2 + 2f_3 h_3 \cos\theta_3 + r_2 \sin\theta_2$$

Solution: $h_3' = 0.10605$

$$f_3' = -3.23968 \text{ in.}$$

(B) Do the kinematic analysis of motions of the centers of mass. (See Fig. 3.8.)

(1) $\quad X_{g2} = r_4 \cos\theta_2 \qquad\qquad\qquad = 2.5981$ in.

$\quad Y_{g2} = r_4 \sin\theta_2 \qquad\qquad\qquad = 1.5000$ in.

$\quad X_{g3} = r_2 \cos\theta_2 \qquad\qquad\qquad = 5.1962$ in.

3.15

$$Y_{g3} = r_2 \sin\theta_2 \qquad\qquad = 3.0000 \text{ in.}$$

$$X_{g4} = r_5 \cos\theta_3 \qquad\qquad = 2.9459 \text{ in.}$$

$$Y_{g4} = -r_1 + r_5 \sin\theta_3 \qquad\qquad = -3.4958 \text{ in.}$$

(2) $\dot{X}_{g2} = (-r_4 \sin\theta_2)\dot\theta_2 \qquad\qquad = -60.00 \text{ in./sec}$

$\dot{Y}_{g2} = (r_4 \cos\theta_2)\dot\theta_2 \qquad\qquad = 103.92 \text{ in./sec}$

$\dot{X}_{g3} = (-r_2 \sin\theta_2)\dot\theta_2 \qquad\qquad = -120.00 \text{ in./sec}$

$\dot{Y}_{g3} = (r_2 \cos\theta_2)\dot\theta_2 \qquad\qquad = 207.85 \text{ in./sec}$

$\dot{X}_{g4} = (-r_5 \sin\theta_3)h_3\dot\theta_2 \qquad\qquad = -97.19 \text{ in./sec}$

$\dot{Y}_{g4} = (r_5 \cos\theta_3)h_3\dot\theta_2 \qquad\qquad = 33.67 \text{ in./sec}$

(3) $\ddot{X}_{g2} = (-r_4 \cos\theta_2)\dot\theta_2^2 \qquad\qquad = -4156.93 \text{ in./sec}^2$

$\ddot{Y}_{g2} = (-r_4 \sin\theta_2)\dot\theta_2^2 \qquad\qquad = -2400.00 \text{ in./sec}^2$

$\ddot{X}_{g3} = (-r_2 \cos\theta_2)\dot\theta_2^2 \qquad\qquad = -8313.86 \text{ in./sec}^2$

$\ddot{Y}_{g3} = (-r_2 \sin\theta_2)\dot\theta_2^2 \qquad\qquad = -4800.00 \text{ in./sec}^2$

$\ddot{X}_{g4} = \left[(-r_5 \sin\theta_3)h_3' - (r_5 \cos\theta_3)h_3^2\right]\dot\theta_2^2 = -1827.73 \text{ in./sec}^2$

$\ddot{Y}_{g4} = \left[(r_5 \cos\theta_3)h_3' - (r_5 \sin\theta_3)h_3^2\right]\dot\theta_2^2 = -610.80 \text{ in./sec}^2$

(C) Make free-body sketches (Fig. 3.9).

Notes: It is contemplated that moments will be taken around "O" on
link 2, around "A" on link 3 and around "C" on link 4. Position
vectors are defined accordingly.

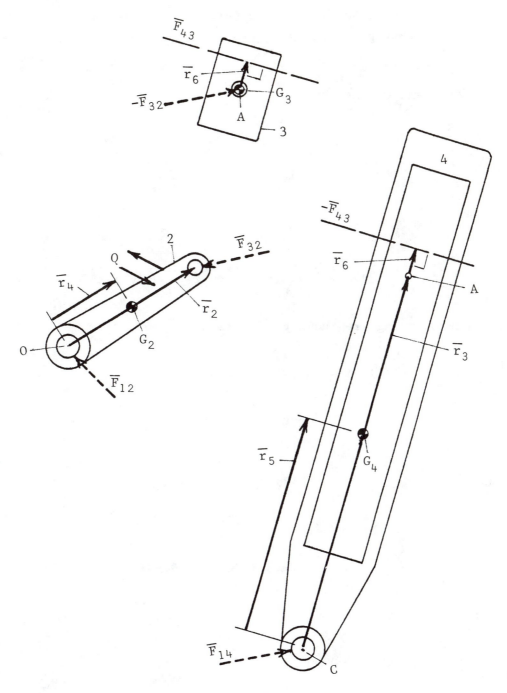

Fig. 3.9 Freebody sketches.

3.17

The location of $\overline{F}_{43}$ is unknown but the direction is known. Vector $\overline{r}_6$ is defined to be perpendicular to the line of action of $\overline{F}_{43}$. The direction angle of the line of action of $\overline{F}_{43}$ is $(\theta_3 + 90 \deg)$.

(D) Write force and moment equations.

(1) Link 2.

$$F_{12x} + F_{32x} = m_2 A_{g2x} \tag{1}$$

$$F_{12y} + F_{32y} = m_2 A_{g2y} \tag{2}$$

$$\Sigma M_0 = Q + r_{2x}F_{32y} - r_{2y}F_{32x} = (r_{4x}A_{g2y} - r_{4y}A_{g2x})m_2 + I_2\alpha_2 \tag{3}$$

The following are known quantities.

$$m_2 A_{g2x} = (3/32.2)(-4156.93/12) = -32.27 \text{ lb}_f$$

$$m_2 A_{g2y} = (3/32.2)(-2400.00/12) = -18.63 \text{ lb}_f$$

$$I_2\alpha_2 = 0$$

$$(r_{4x}A_{g2y} - r_{4y}A_{g2x})m_2 = [3\cos30°(-2400.00/12)$$

$$- 3\sin30°(-4156.93/12)](3/32.2) = 0$$

$$r_{2x} = 6\cos30° = 5.1962 \text{ in.}$$

$$r_{2y} = 6\sin30° = 3.0000 \text{ in.}$$

So, for link 2,

$$F_{12x} + F_{32x} = -32.37 \text{ lb}_f \tag{4}$$

3.18

$$F_{12y} + F_{32y} = -18.63 \text{ lb}_f \tag{5}$$

$$Q + 5.1962 \, F_{32y} - 3.0000 \, F_{32x} = 0 \tag{6}$$

(2) Link 3.

$$F_{43x} + (-F_{32x}) = m_3 A_{g3x} \tag{7}$$

$$F_{43y} + (-F_{32y}) = m_3 A_{g3y} \tag{8}$$

$$\Sigma M_A = r_{6x} F_{43y} - r_{6y} F_{43x} = I_3 \alpha_3 \tag{9}$$

But,

$$m_3 A_{g3x} = (2/32.2)(-8313.86/12) = -43.03 \text{ lb}_f$$

$$m_3 A_{g3y} = (2/32.2)(-4800.00/12) = -24.84 \text{ lb}_f$$

$$\alpha_3 = h_3' \ddot{\theta}_2^2 = (0.10605)(40)^2 = 169.68 \text{ rad/sec}^2$$

$$I_3 \alpha_3 = \frac{0.5(169.68)}{(32.2)(12)} = 0.22 \text{ in.-lb}_f$$

$$F_{43x} = F_{43}\cos(\theta_3 + 90°) = F_{43}\cos 160.984° = -0.9449 F_{43}$$

$$F_{43y} = F_{43}\sin 160.894° = 0.3273 F_{43}$$

Since $\overline{r}_6$ was defined to be perpendicular to $\overline{F}_{43}$,

$$r_{6x} F_{43y} - r_{6y} F_{43x} = r_6 F_{43}$$

With these substitutions, equations (7)-(9) become

3.19

$$-0.9449F_{43} - F_{32x} = -43.03 \tag{10}$$

$$0.3273F_{43} - F_{32y} = -24.84 \tag{11}$$

$$r_6F_{43} = 0.22 \text{ in.-lb}_f \tag{12}$$

(3) Link 4.

$$-F_{43x} + F_{14x} = m_4A_{g4x} \tag{13}$$

$$-F_{43y} + F_{14y} = m_4A_{g4y} \tag{14}$$

$$\Sigma M_c = (r_3 + r_6)(-F_{43}) = (r_{5x}A_{g4y} - r_{5y}A_{g4x})m_4 + I_4\alpha_4 \tag{15}$$

where,

$$r_3 = 15.875 \text{ in.}$$

$$m_4A_{g4x} = (8/32.2)(-1827.73/12) = -37.84 \text{ lb}_f$$

$$m_4A_{g4y} = (8/32.2)(-610.80/12) = -12.65 \text{ lb}_f$$

$$(r_{5x}A_{g4y} - r_{5y}A_{g4x})m_4 = [(9\cos70.894°)(-610.80/12)$$

$$- (9\sin70.894°)(-1827.73/12)](8/32.2)$$

$$= 284.56 \text{ in.-lb}_f$$

$$\alpha_4 = \alpha_3 = 169.68 \text{ rad/sec}^2$$

$$I_4\alpha_4 = \frac{(200)(169.68)}{(32.2)(12)} = 87.83 \text{ in.-lb}_f$$

3.20

Then,

$$0.9449F_{43} + F_{14x} = -37.84 \tag{16}$$

$$-0.3273F_{43} + F_{14y} = -12.65 \tag{17}$$

$$-15.874F_{43} - r_6F_{43} = 372.39 \tag{18}$$

(E) Solve

Equation (12): $r_6F_{43} = 0.22$

Equation (18): $-15.875F_{43} - 0.22 = 372.39$

$$F_{43} = -23.47 \text{ lb}_f$$

$$r_6 = -0.0094 \text{ in.}$$

Equation (16): $0.9449(-23.47) + F_{14x} = -37.84$

$$F_{14x} = -15.66 \text{ lb}_f$$

Equation (17): $-0.3273(-23.47) + F_{14y} = -12.65$

$$F_{14y} = -20.33 \text{ lb}_f$$

Equation (10): $-0.9449(-23.47) - F_{32x} = -43.03$

$$F_{32x} = 65.21 \text{ lb}_f$$

Equation (11): $0.3273(-23.47) - F_{32y} = -24.84$

$$F_{32y} = 17.16 \text{ lb}_f$$

Equation (4): $F_{12x} + 65.21 = -32.27$

$$F_{12x} = -97.48 \text{ lb}_f$$

Equation (5): $F_{12y} + 17.16 = -18.63$

$$F_{12y} = -35.79 \text{ lb}_f$$

Equation (6): $Q + 5.1962(17.16) - 3.0000(65.21) = 0$

$$Q = 106.5 \text{ in.-lb}_f$$

(F)   Absolute values and directions of forces.   Results are pictured in Fig. 3.10.

$$F_{12} = \sqrt{F_{12x}^2 + F_{12y}^2} = \sqrt{(-97.48)^2 + (-35.79)^2}$$

$$= 103.8 \text{ lb}_f \text{ at } 200.2 \text{ deg.}$$

$$F = \sqrt{F_{32x}^2 + F_{32y}^2} = \sqrt{(65.21)^2 + (17.16)^2}$$

$$= 67.4 \text{ lb}_f \text{ at } 14.7 \text{ deg.}$$

$$F_{43} = 23.5 \text{ lb}_f \text{ at } -19.1 \text{ deg.}$$

$$F_{14} = \sqrt{F_{14x}^2 + F_{14y}^2} = \sqrt{(-15.66)^2 + (-20.33)^2}$$

$$= 25.7 \text{ lb}_f \text{ at } 232.4 \text{ deg.}$$

3.6   Matrix Arrangement of Force and Moment Equations

In the example of the previous article, after substituting known information from the kinematic analysis, we had nine linear equations in nine unknowns
to solve. We carried out the solution a step at a time, which is an appropriate
process for hand calculation for one position. For automated computation some

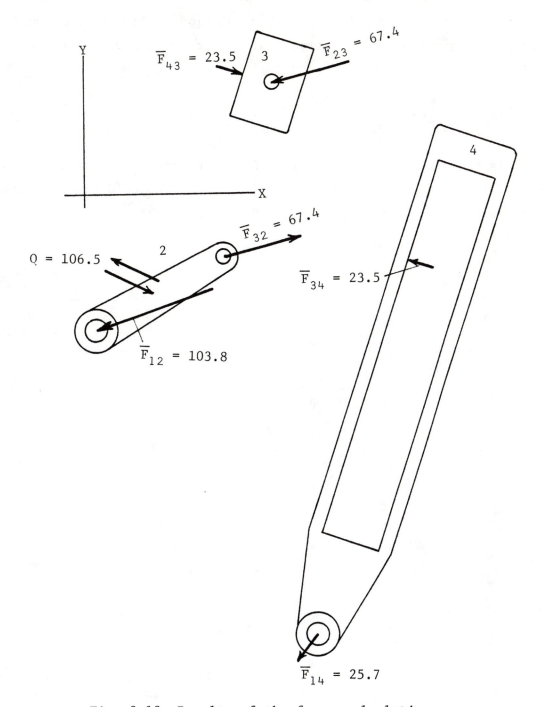

Fig. 3.10   Results of the force calculations.

people prefer to treat the problem as one of solving nine linear equations simultaneously using a general program for solution of linear equations. A first step in organizing the solution would be to express the nine equations in matrix form

$$[A]\{F\} = \{B\} \tag{1}$$

where $\{F\}$ is the vector (column matrix) of nine unknowns, $\{B\}$ is the vector (column matrix) of nine known quantities from the right-hand sides of the equations and $[A]$ is the 9×9 matrix of coefficients of the unknowns.

The nine equations were

$$F_{12x} + F_{32x} = -32.37$$

$$F_{12y} + F_{32y} = -18.63$$

$$-3F_{32x} \quad 5.1962F_{32y} + Q = 0$$

$$-F_{32x} \quad -0.9449F_{43} = -43.03$$

$$-F_{32y} \quad 0.3273F_{43} = -24.84$$

$$r_6F_{43} = 0.22$$

$$0.9449F_{43} + F_{14x} = -37.84$$

$$-0.3273F_{43} \quad +F_{14y} = -12.65$$

$$-15.875F_{43} - r_6F_{43} = 372.39$$

The column matrix of unknowns is

$$\{F\} = \begin{bmatrix} F_{12x} \\ F_{32x} \\ F_{12y} \\ F_{32y} \\ Q \\ F_{43} \\ r_6 F_{43} \\ F_{14x} \\ F_{14y} \end{bmatrix}$$

The corresponding column matrix of known terms is

$$\{B\} = \begin{bmatrix} -32.37 \\ -18.63 \\ 0 \\ -43.03 \\ -24.84 \\ 0.22 \\ -37.84 \\ -12.65 \\ 372.39 \end{bmatrix}$$

The matrix of coefficients is

$$[A] = \begin{bmatrix} 1 & 1 & 0 & 0 & 0 & 0 & 0 & 0 & 0 \\ 0 & 0 & 1 & 1 & 0 & 0 & 0 & 0 & 0 \\ 0 & -3 & 0 & 5.1962 & 1 & 0 & 0 & 0 & 0 \\ 0 & -1 & 0 & 0 & 0 & -0.9449 & 0 & 0 & 0 \\ 0 & 0 & 0 & -1 & 0 & 0.3273 & 0 & 0 & 0 \\ 0 & 0 & 0 & 0 & 0 & 0.9449 & 0 & 1 & 0 \\ 0 & 0 & 0 & 0 & 0 & 0 & 0 & 1 & 0 \\ 0 & 0 & 0 & 0 & 0 & -0.3273 & 0 & 0 & 1 \\ 0 & 0 & 0 & 0 & 0 & -15.875 & -1 & 0 & 0 \end{bmatrix}$$

3.7 Three-Dimensional Aspects of Plane Motion Force Analysis

Even though we limit ourselves to plane motion mechanisms, the force problem is not completely two-dimensional. The moving masses will not be concentrated in one plane. Also the force lines may fall in different planes. Consider, for example, the possible construction and support of link 2 of the mechanism in Fig. 3.6. The arrangement might be as pictured in Fig. 3.11. If so, then the force $\overline{F}_{12}$ which we calculated is really the resultant of the reactions at the bearings B and D. It is these individual reactions that we should know for design purposes. We will need to supplement the results of our two-dimensional force analysis.

Consider the object pictured in Fig. 3.12. Motion is two-dimensional, parallel to the X,Y plane. The object has a plane of symmetry which is parallel to the X,Y plane. For this case the force and moment equations are

$$\Sigma F_x = mA_{gx} \tag{1}$$

$$\Sigma F_y = mA_{gy} \tag{2}$$

$$\Sigma M_z = m(r_x A_{gy} - r_y A_{gx}) + I_g \alpha \tag{3}$$

$$\Sigma F_z = 0 \tag{4}$$

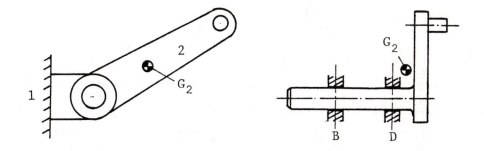

Fig. 3.11   Three-dimensional nature of the crank from Fig. 3.6.

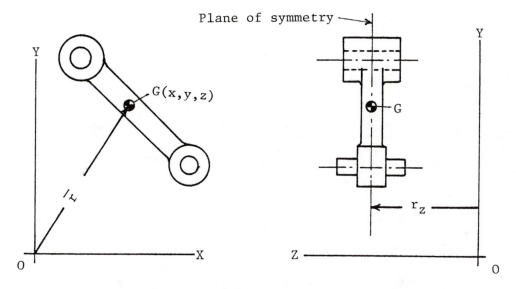

Fig. 3.12   Part having a plane of symmetry parallel to the plane
of motion (X,Y plane).

$$\Sigma M_x = -mr_z A_{gy} \tag{5}$$

$$\Sigma M_y = mr_z A_{gx} \tag{6}$$

Equations (1)-(3) are the two-dimensional equations. Equations (4)-(6) are the supplementary equations needed to account for the three-dimensional aspects of the problem. Equations (1)-(4) apply regardless of the shape of the object, but equations (5) and (6) are correct only if the object has a plane of symmetry parallel to the plane of motion.

Many machine parts will not possess the symmetry required for equations (5) and (6), but may be composites of shapes which individually have the described symmetry. See, for example, Fig. 3.13. In this case, equations (5) and (6) must be modified as follows.

$$\Sigma M_x = -\Sigma mr_z A_{gy} \tag{7}$$

$$\Sigma M_y = \Sigma mr_z A_{gx} \tag{8}$$

The right-hand sides are to be evaluated by summing the terms for each symmetrical piece of the composite shape.

3.8 Example. Bearing Reactions on Crank Shaft

The equations for a two-dimensional force analysis of the crank shown in Fig. 3.14 are

$$P_x + R_x = mA_{gx} \tag{1}$$

$$P_y + R_y = mA_{gy} \tag{2}$$

$$Q + (r_{px}P_y - r_{py}P_x) = m(r_{gx}A_{gy} - r_{gy}A_{gx}) + I_g\alpha \tag{3}$$

3.28

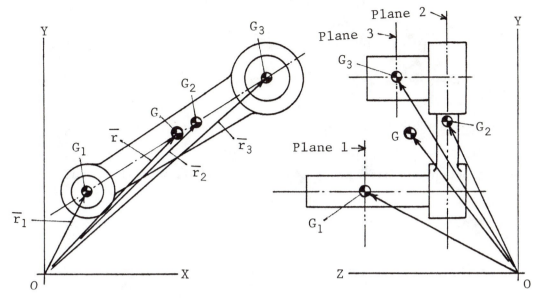

Fig. 3.13   A machine part which does not have a plane of symmetry parallel to the plane of motion (X,Y plane) but is a composite of 3 shapes each individually having such symmetry.

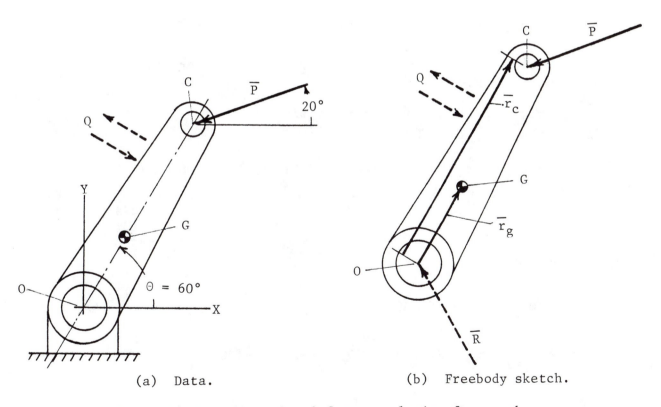

(a)   Data.   (b)   Freebody sketch.

Fig. 3.14   Two-dimensional force analysis of a crank.

From the given data,

$$P_x = -69\cos 20° = -65 \text{ lb}_f$$

$$P_y = -69\sin 20° = -24 \text{ lb}_f$$

$$r_{px} = 6\cos 60° = 3.00 \text{ in.,} \qquad r_{gx} = 2.32\cos 60° = 1.16 \text{ in.}$$

$$r_{py} = 6\sin 60° = 5.20 \text{ in.,} \qquad r_{gy} = 2.32\sin 60° = 2.01 \text{ in.}$$

$$m = 7.6/32.2 = 0.236 \text{ slugs}$$

$$I_g = 100.4/(386) = 0.260 \text{ slug-ft-in.}$$

$$A_{gx} = \frac{d^2}{dt^2}(r_g\cos\theta) = -r_g\sin\theta\,\ddot{\theta} - r_g\cos\theta\,\dot{\theta}^2$$

$$= -r_{gy}\alpha - r_{gx}\omega^2 = -2.01(800) - 1.16(60)^2 = -5784 \text{ in./sec}^2$$

$$(\text{or } -482 \text{ ft/sec}^2)$$

$$A_{gy} = \frac{d^2}{dt^2}(r_g\sin\theta) = r_g\cos\theta\,\ddot{\theta} - r_g\sin\theta\,\dot{\theta}^2$$

$$= r_{gx}\alpha - r_{gy}\omega^2 = 1.16(800) - 2.01(60)^2 = -6308 \text{ in./sec}^2$$

$$(\text{or } -526 \text{ ft/sec}^2)$$

From equations (1)-(3)

$$R_x = mA_{gx} - P_x = (0.236)(-482) - (-65) = -49 \text{ lb}_f$$

$$R_y = mA_{gy} = P_y = (0.236)(-526) - (-24) = -100 \text{ lb}_f$$

$$Q = m(r_{gx} A_{gy} - r_{gy} A_{gx}) + I_g \alpha - r_{px} P_y + r_{py} P_x$$

$$= (0.236)(1.16)(-526) - (2.01)(-482) + (0.260)(800)$$

$$- (3.00)(-24) + (5.20)(-65)$$

$$= 27 \text{ lb}_f \text{ in.}$$

This completes the two-dimensional analysis. We now know the driving torque, Q, and the resultant of the reactions at the fixed bearings. To find individual bearing reactions we need another view of the crank to show how the mass is distributed perpendicular to the plane of motion, and where the bearings are located. See Fig. 3.15 for this additional information.

The complete crank is a composite of three pieces (pin, crankbody, shaft) each having a plane of symmetry parallel to the X,Y plane (the plane of motion). To find the reactions $\overline{R}_A$ and $\overline{R}_B$ at bearings A and B we need to solve the moment equations around the X and Y axes.

$$-Z_a R_{ay} - Z_1 P_y = -m_1 Z_1 A_{g1y} - m_2 Z_2 A_{g2y} - m_3 Z_3 A_{g3y} \qquad (4)$$

$$Z_a R_{ax} + Z_1 P_x = m_1 Z_1 A_{g1x} + m_2 A_2 A_{g2x} + m_3 Z_3 A_{g3x} \qquad (5)$$

$$Z_a = 4 \text{ in.} \qquad A_1 = -3.5 \text{ in.} \qquad Z_2 = -2 \text{ in.}$$

$$Z_3 = 3 \text{ in.}$$

$$A_{g1x} = -1246 \text{ ft/sec}^2 \qquad\qquad A_{g1y} = -1359 \text{ ft/sec}^2$$

$$A_{g2x} = -582 \text{ ft/sec}^2 \qquad\qquad A_{g2y} = -634 \text{ ft/sec}^2$$

$$A_{g3x} = 0 \text{ ft/sec}^2 \qquad\qquad A_{g3y} = 0 \text{ ft/sec}^2$$

$$m_1 = 0.019 \text{ slugs}, \quad m_2 = 0.155 \text{ slugs}, \quad m_3 = 0.062 \text{ slugs}$$

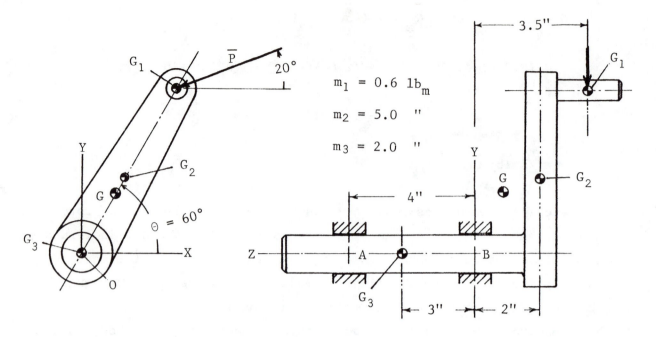

$m_1 = 0.6 \ \text{lb}_m$

$m_2 = 5.0 \quad "$

$m_3 = 2.0 \quad "$

Fig. 3.15 Additional data needed to solve for bearing reactions at A and B.

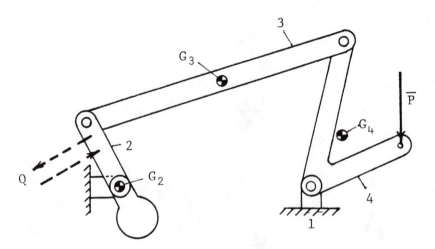

Fig. 3.16 Example to illustrate separation of total force analysis into static and inertia analyses. External load P. Driving couple Q applied to link 2. See Fig. 3.17 for freebody sketches.

$$-4R_{ay} - (-3.5)(-24) = -(0.019)(-3.5)(-1359) - (0.155)(-2)(-634)$$

$$- (0.062)(3)(0)$$

$$R_{ay} = 51 \text{ lb}_f$$

$$4R_{ax} + (-3.5)(-65) = (0.019)(-3.5)(-1246) + (0.155)(-2)(-582)$$

$$+ (0.062)(3)(0)$$

$$R_{ax} = 9 \text{ lb}_f$$

We now solve for the reaction components at bearing B.

$$R_{bx} = R_x - R_{ax} = -49 - 9 = -58 \text{ lb}_f$$

$$R_{by} = R_y - R_{ay} = -100 - 51 = -151 \text{ lb}_f$$

The resultant forces on the shaft at bearings A and B are

$$R_a = \sqrt{(9)^2 + (51)^2} = 52 \text{ lb}_f \text{ at 80 deg.}$$

$$R_b = \sqrt{(-58)^2 + (-151)^2} = 162 \text{ lb}_f \text{ at 69 deg.}$$

3.9 Static Force Analysis. Inertia Force Analysis

The complete force analysis of a mechanism is sometimes broken into two parts, a static analysis and an inertia analysis. In the static analysis the parts of the mechanism are treated as though they had no mass. External loads on the mechanism are taken into account and the resulting joint forces determined. In the inertia analysis the masses are taken into account and the external loadings are ignored. If friction is neglected, then the results of the two analyses can be superposed to give the complete force results.

Consider the 4-bar linkage shown in Fig. 3.16. Force $\overline{P}$ on link 4 is the external load. The mechanism is driven by couple Q applied to link 2. For a

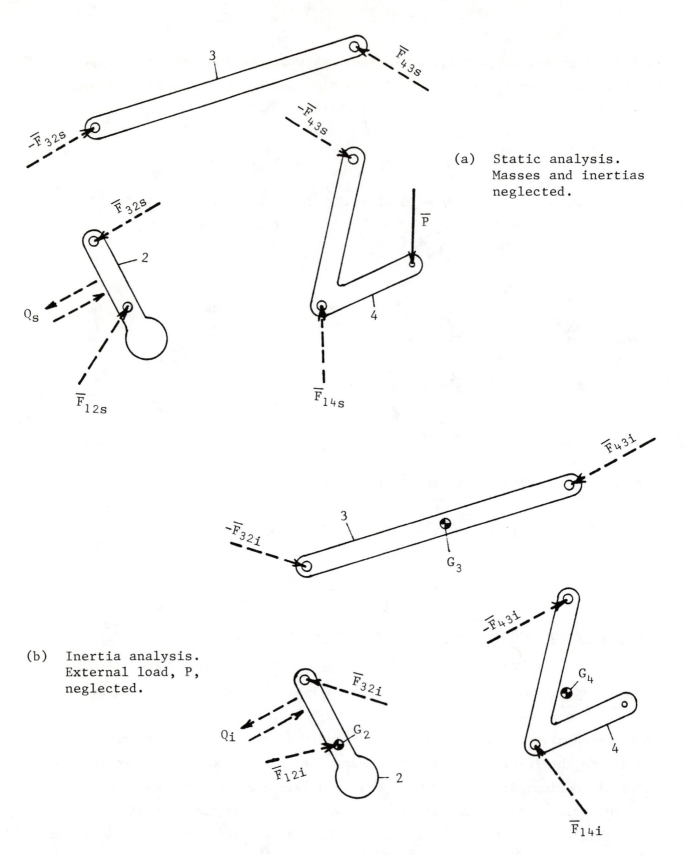

(a) Static analysis. Masses and inertias neglected.

(b) Inertia analysis. External load, P, neglected.

Fig. 3.17 Freebody sketches.

3.34

static force analysis the free-body sketches would be as shown in Fig. 3.17(a). For the inertia analysis we would ignore the load $\overline{P}$, but include the inertia terms in the equations. The free-body sketches would be as indicated in Fig. 3.17(b).

After making the two solutions we could then say,

$$Q = Q_s + Q_i$$

$$(\overline{F}_{32})_{TOT} = (\overline{F}_{32})_{STATIC} + (\overline{F}_{32})_{INERTIA}$$

$$(\overline{F}_{14})_{TOT} = (\overline{F}_{14})_{STATIC} + (\overline{F}_{14})_{INERTIA} \qquad \text{etc.}$$

E3.1 $\overline{P}$ is a known force on link 8. Q is an unknown couple on link 2. Define
suitable vectors for a kinematic analysis then assume that the basic loop
analysis has been completed.

Complete the freebody sketches for a static force analysis of the mechanism. Write all force and moment equations. Carefully define all symbols used.

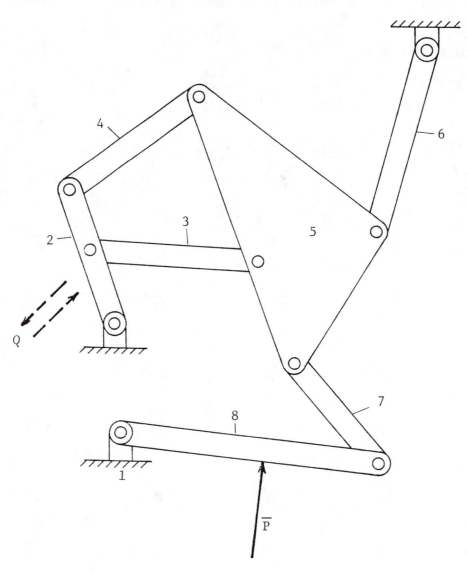

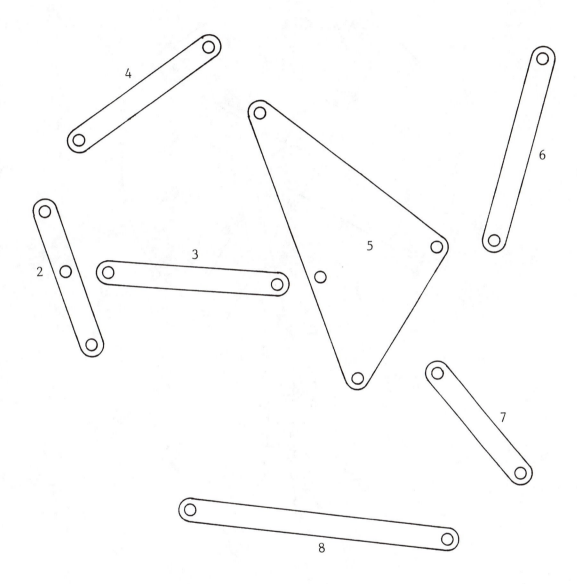

E3.2 Make freebody sketches of the moving parts of the mechanism. Carefully define all symbols used. Write force and moment equations for a static force analysis. Assume that the basic vector loop kinematic analysis has already been completed.

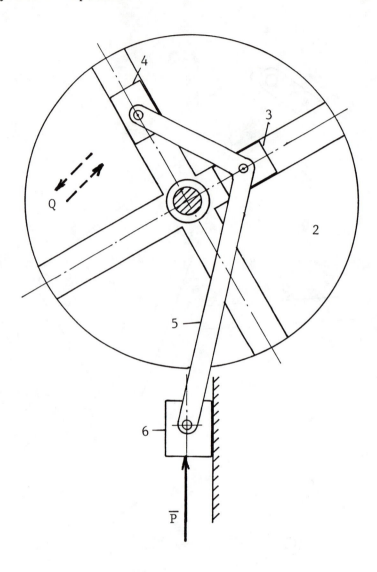

E3.3 Complete the freebody sketches for an inertia force analysis of the mechanism. Assume couple Q to be driving. Assume crank 2 is rotating at constant speed. Also assume that the basic vector loop kinematic analysis has been completed for the position to be considered. Write all force and moment equations. Carefully define all symbols used.

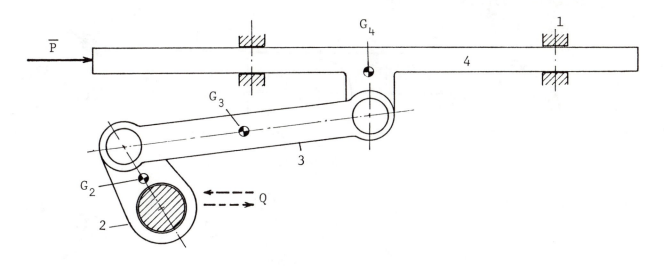

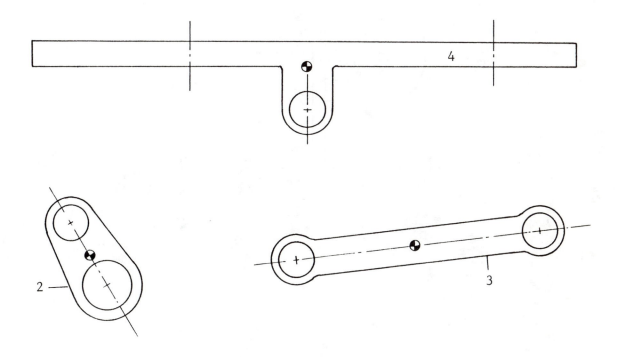

E3.4 Work up a program for the complete force analysis of this mechanism for
      a full rotation of the driving crank. $\overline{P}$ is a known force on 4 and Q is
      the unknown driving couple on crank 2. All geometric and physical con-
      stants of the links are known. Input crank 2 is turning counter-clock-
      wise with a constant angular velocity. Your program should include the
      necessary kinematic computations.

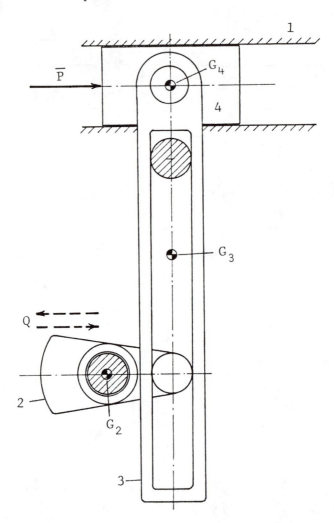

E3.4 Continued.

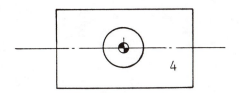

4

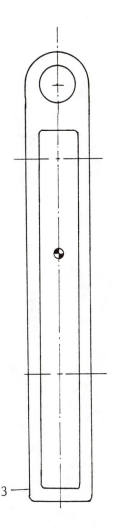

2

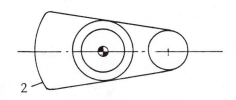

3

CHAPTER 4

FORCE ANALYSIS II

4.1 Introduction

In the previous chapter we looked at the force analysis of plane mechanisms and discussed some examples in which friction was ignored. For many situations this would be quite adequate. In machinery that runs at moderate to high speed, with good bearings and/or well lubricated joints, the forces predicted by an analysis ignoring friction will ordinarily be quite satisfactory for design purposes.

For devices which operate at low speed and/or with relatively poor bearings, friction effects are likely to be more important. In this chapter we will look at ways of dealing with Coulomb friction in joints of a mechanism.

4.2 Pin Joints

The situation within a pin joint is pictured in the series of sketches of Fig. 4.1. The "normal" load is distributed in some fashion, never really known, over a portion of the joint surface (Fig. 4.1(b)). The corresponding elementary friction forces are then distributed tangent to the joint surface. In view of the uncertainty in the distribution of load, and of the value of the friction coefficient, it is usually considered adequate to assume point loading, as shown in Fig. 4.1(c). Note that, if μ = friction coefficient and R = pin radius, then $\tan\phi = \mu$ and $R\sin\phi = R\mu/\sqrt{1 + \mu^2}$ = radius of the small circle to which $\overline{F}_{32}$ is

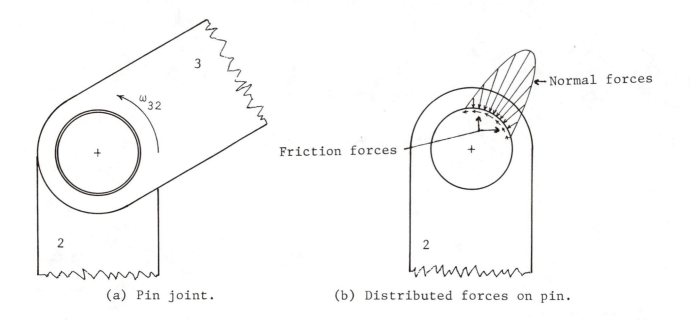

(a) Pin joint. (b) Distributed forces on pin.

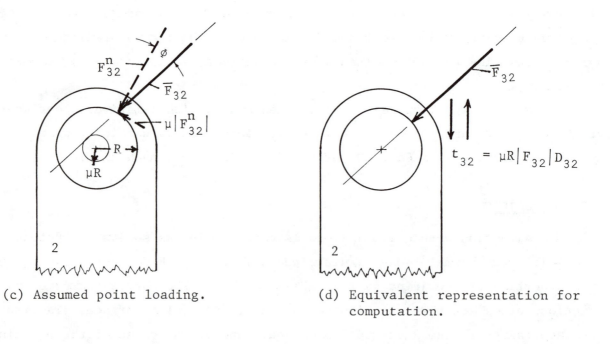

(c) Assumed point loading. (d) Equivalent representation for computation.

Fig. 4.1 Force situation in a pin joint with Coulomb friction.

tangent. In view of the fact that μ typically is 0.1 or less, the factor $\sqrt{1 + \mu^2}$ is very close to unity and is so assumed. Hence we say that the small circle, called the "friction circle," has radius "μR." Note that the force $\overline{F}_{32}$ is tangent to this circle on such a side of the pin center that the moment of $\overline{F}_{32}$ around the pin center is in the sense of ω_{32}. The representation in Fig. 4.1(c) with the resultant force $\overline{F}_{32}$ drawn tangent to the friction circle is a good one for graphical force analysis. For a computerized analysis the equivalent representation in Fig. 4.1(d) is much easier to work with. Here the resultant force $\overline{F}_{32}$ is replaced by an equal force through the pin center plus a couple, t_{32}, equal to the moment of $\overline{F}_{32}$ with respect to the pin center.

$$t_{32} = \mu R \left| F_{32} \right| D_{32}$$

where D_{32} is a direction indicator defined to be +1 if ω_{32} is counterclockwise and -1 if ω_{32} is clockwise.

4.3 The Computation Problem

The main problem introduced by Coulomb friction is that our set of force and moment equations will not be linear. It is suggested that we deal with this by an iterative solution, as follows:

(1) Write all force and moment equations, including the appropriate friction terms.

(2) Set all friction terms equal to zero and solve for the forces.

(3) Use the results of step (2) to evaluate the friction terms, substitute into the equations and solve again.

(4) Re-evaluate the friction terms based on the results of step (3) and repeat until satisfactory accuracy is obtained.

This procedure will be illustrated in the examples to follow.

4.4 Simple Example (Fig. 4.2)

The object pictured is being turned clockwise by force $\overline{P}$. Weight $\overline{W}$ resists the motion. The friction coefficient in the bearing is taken to be 0.2. Inertia

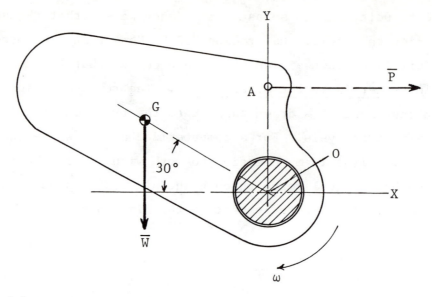

(a) Data. OG = 8 in. OA = 6 in. Bearing radius = 2 in.
W = 30 lb. Friction coefficient = 0.2.

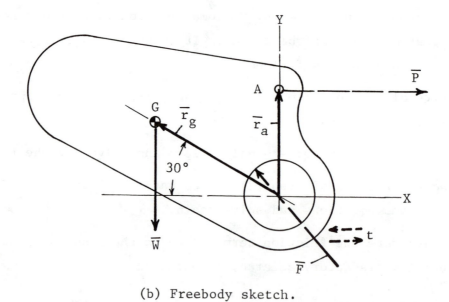

(b) Freebody sketch.

Fig. 4.2 Simple example to illustrate handling of pin friction.

4.4

is to be neglected. The equations to be solved are:

$$P_x + W_x + F_x = 0 \tag{1}$$

$$P_y + W_y + F_y = 0 \tag{2}$$

$$r_{gx}W_y - r_{gy}W_x + r_{ax}P_y - r_{ay}P_x + t = 0 \tag{3}$$

But

$$P_x = P, \qquad P_y = 0, \qquad W_x = 0, \qquad W_y = -30 \ \text{lb}_f$$

$$r_{gx} = 8 \cos 150° = -6.93 \ \text{in.}, \qquad r_{gy} = 8 \sin 30° = 4.00 \ \text{in.}$$

$$r_{ax} = 0, \qquad r_{ay} = 6.00 \ \text{in.}$$

So the equations become

$$P + F_x = 0 \tag{4}$$

$$-W + F_y = 0 \tag{5}$$

$$207.8 - 6.00P + t = 0 \tag{6}$$

Solution:

(1) Set $t = 0$ and solve equations (4)-(6).

$$P = 207.8/6 = 34.63$$

$$F_x = -34.63, \qquad F_y = 30$$

(2) Evaluate t

$$|F| = \sqrt{(34.63)^2 + (30)^2} = 45.82$$

$$t = \mu R |F| D = (0.2)(2)(45.82)(+1) = 18.3 \ \text{in.lb}_f$$

(3) Substitute this value of t into the equations and re-solve.

$$P = (207.8 + 18.3)/6 = 37.68$$

$$F_x = -37.68, \qquad F_y = 30$$

(4) Re-evaluate t

$$|F| = \sqrt{(37.68)^2 + (30)^2} = 48.16$$

$$t = (0.2)(2)(48.16)(+1) = 19.3$$

(5) Substitute this value into the equations and solve again.

$$P = (207.8 + 19.3)/6 = 37.85$$

$$F_x = -37.85, \qquad F_y = 30, \qquad |F| = 48.3$$

It is obvious that continuing the process is not going to affect the results materially. We can accept

$$P = 37.9 \text{ lb}_f, \qquad F = 48.3 \text{ lb}_f$$

If we had stopped at the end of step (3) we would have been off by only 0.5% approximately.

The procedure illustrated by this elementary example can be followed in programming the solution for a mechanism. The difference will be that there will be more equations to solve and more joints to consider.

4.5 Four-Bar Example (Fig. 4.3)

We will outline the static force analysis for the 4-bar mechanism in the position shown ($\theta_2 = 100$ deg.). We have given the necessary constants of the mechanism, and the fact that force $\overline{P}$ is driving. Q is a resisting couple, given, applied to link 2. The friction radius, μR, is 0.125 in. for each joint.

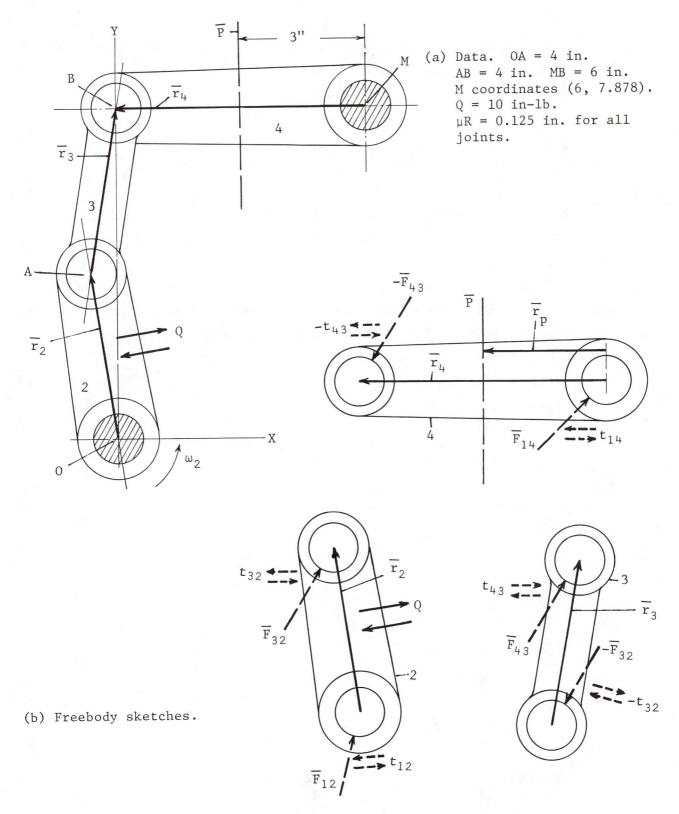

(a) Data. OA = 4 in.
AB = 4 in. MB = 6 in.
M coordinates (6, 7.878).
Q = 10 in-lb.
μR = 0.125 in. for all
joints.

(b) Freebody sketches.

Fig. 4.3 Data and freebody sketches for problem of Art. 4.5.

4.7

We also assume that the kinematic analysis has already been done and that we have the results, in particular, at $\Theta_2 = 100$ deg.,

$$\Theta_3 = 80°, \qquad \Theta_4 = 180°$$

$$h_3 (= d\Theta_3/d\Theta_2) = -1.000, \qquad h_4 (= d\Theta_4/d\Theta_2) = 0.232$$

The free-body diagrams are shown in Fig. 4.3(b). Note that at each joint we have identified the appropriate force and friction couple. For example, on link 2 at A the force is $\overline{F}_{32}$ and the friction couple t_{32}. The broken vectors indicate that these are unknowns. Although we have placed arrowheads on the force vectors and have shown a direction for the couple, this does not mean we have any prior knowledge or have made any assumption about directions. All we have really done is to name these items.

The equations to be solved are

Forces, Link 2

$$F_{12x} + F_{32x} = 0 \tag{1}$$

$$F_{12y} + F_{32y} = 0 \tag{2}$$

Forces, Link 3

$$-F_{32x} + F_{43x} = 0 \tag{3}$$

$$-F_{32y} + F_{43y} = 0 \tag{4}$$

Forces, Link 4

$$-F_{43x} + F_{14x} + P_x = 0 \tag{5}$$

$$-F_{43y} + F_{14y} + P_y = 0 \tag{6}$$

Moments, Link 2

$$Q + r_{2x}F_{32y} - r_{2y}F_{32x} + t_{32} + t_{12} = 0 \tag{7}$$

Moments, Link 3

$$r_{3x}F_{43y} - r_{3y}F_{43x} - t_{32} + t_{43} = 0 \tag{8}$$

Moments, Link 4

$$r_{4x}(-F_{43y}) - r_{4y}(-F_{43x}) + r_{px}P_y - r_{py}P_x - t_{43} + t_{14} = 0 \tag{9}$$

Substitutions of known data:

$$P_x = 0, \qquad P_y = P, \qquad Q = -10 \text{ in.} 1b_f$$

$$r_{2x} = 4 \cos 100° = -0.695, \qquad r_{2y} = 4 \sin 100° = 3.939$$

$$r_{3x} = 4 \cos 80° = 0.695, \qquad r_{3y} = 4 \sin 80° = 3.939$$

$$r_{4x} = 6 \cos 180° = -6.000, \qquad r_{4y} = 4 \sin 180° = 0$$

$$r_{px} = 3 \cos 180° = -3.000, \qquad r_{py} = 3 \sin 180° = 0$$

Then equations (1)-(9) become

$$F_{12x} + F_{32x} = 0 \tag{10}$$

$$F_{12y} + F_{32y} = 0 \tag{11}$$

$$-F_{32x} + F_{43x} = 0 \tag{12}$$

$$-F_{32y} + F_{43y} = 0 \tag{13}$$

$$-F_{43y} + F_{14x} = 0 \tag{14}$$

$$-F_{43y} + F_{14y} + P = 0 \tag{15}$$

$$-0.695F_{32y} - 3.939F_{32x} = 10 - t_{32} - t_{12} \tag{16}$$

$$0.695F_{43y} - 3.939F_{43x} = t_{32} - t_{43} \tag{17}$$

$$6.000F_{43y} - 3.000P = t_{43} - t_{14} \tag{18}$$

Solution procedure:

(1) We set all the friction couples equal to zero and solve equations (10)-(18). The results are shown in column (1) of the table in Fig. 4.4.

(2) Next we evaluate the friction couples using the forces determined in step (1). For example,

$$t_{32} = \mu R |F_{32}| D_{32}$$

$$|F_{32}| = \sqrt{F_{32x}^2 + F_{32y}^2} = \sqrt{(1.27)^2 + (7.20)^2} = 7.31$$

The direction indicator, D_{32}, is to be +1 if $\omega_{3/2}$ is counterclockwise and −1 if $\omega_{3/2}$ is clockwise. From the results of the kinematic analysis we can determine D_{32} as follows.

$$D_{32} = \frac{\omega_{3/2}}{|\omega_{3/2}|} = \frac{\omega_3 - \omega_2}{|\omega_3 - \omega_2|} = \frac{\omega_2(h_3 - 1)}{|\omega_2(h_3 - 1)|}$$

$$D_{32} = \frac{\omega_2}{|\omega_2|} \frac{(h_{3-1})}{|h_{3-1}|} = D_{21} \frac{h_{3-1}}{|h_{3-1}|}$$

where D_{21} = +1 if ω_2 is counterclockwise and −1 if ω_2 is clockwise. In our example, ω_2 is counterclockwise, so

$$D_{32} = (+1) \frac{(-1.000 - 1)}{|-1.000 - 1|} = -1$$

Finally,

$$t_{32} = (0.125)(7.31)(-1) = -0.91$$

(3) After evaluating all the friction couples, substitute and re-solve equations (10)-(18). The results are shown in the bottom portion of column (2) in table Fig. 4.4. Continuing the iterative process gives the results shown in the succeeding columns of Fig. 4.4. The convergence is somewhat slow. This is because the example has been deliberately chosen to exaggerate the effect of friction. The mechanism is close to a dead center position and the friction circle radius is typical of poorly lubricated bearings.

4.6 Efficiency of Mechanism

In the example of the previous article the magnitude of the power input is equal to $\overline{P} \cdot \overline{V}_p$ where $\overline{V}_p$ is the velocity of link 4 at the point where driving force $\overline{P}$ is applied.

$$\overline{V}_p = \frac{d}{dt}\left(\overline{r}_p\right) = \frac{d}{dt}\left(r_p \cos \Theta_4 \; \overline{i} + r_p \sin \Theta_4 \; \overline{j}\right)$$

$$= -r_p \sin \Theta_4 \; \dot{\Theta}_4 \; \overline{i} + r_p \cos \Theta_4 \; \dot{\Theta}_4 \; \overline{j}$$

$$= 0 + 3(-1) \; \dot{\Theta}_4 \; \overline{j}$$

$$\overline{P} = -23.43 \; \overline{j}$$

$$\overline{P} \cdot \overline{V}_p = (-23.43 \; \overline{j}) \cdot (-3 \; \dot{\Theta}_4 \; \overline{j}) = 70.3 \; \dot{\Theta}_4$$

The magnitude of the power output is $\left|Q \; \dot{\Theta}_2\right|$

$$= (10)(\dot{\Theta}_2)$$

Hence the instantaneous efficiency is

$$\text{Eff} = \text{Power out/Power in} = (10 \; \dot{\Theta}_2)/(70.3 \; \dot{\Theta}_4)$$

$$= (10/70.3)(1/h_4) = (10/70.3)(1/0.232) = 0.613 \text{ or } 61.3\%$$

	(1)	(2)	(3)	(4)	(5)	(6)
t_{12}	0	− 0.91	− 1.24	− 1.36	− 1.40	− 1.41
t_{32}	0	− 0.91	− 1.24	− 1.36	− 1.40	− 1.41
t_{43}	0	0.91	1.24	1.36	1.40	1.41
t_{14}	0	− 0.91	− 1.31	− 1.46	− 1.52	− 1.53
Q	−10	−10	−10	−10	−10	
F_{12x}	1.27	1.27	1.27	1.27	1.27	
F_{12y}	7.20	9.82	10.77	11.11	11.23	
F_{32x}	− 1.27	− 1.27	− 1.27	− 1.27	− 1.27	
F_{32y}	− 7.20	− 9.82	−10.77	−11.11	−11.23	
F_{43x}	− 1.27	− 1.27	− 1.27	− 1.27	− 1.27	
F_{43y}	− 7.20	− 9.82	−10.77	−11.11	−11.23	
F_{14x}	− 1.27	− 1.27	− 1.27	− 1.27	− 1.27	
F_{14y}	7.20	10.43	11.62	12.05	12.20	
P	−14.40	−20.25	−22.39	−23.16	−23.43	

Fig. 4.4   Table showing progress of computations for the example of Fig. 4.3. Units are lb_f and inches.

Notice that this is the same (within computational accuracy) as the ratio of P_o/P, where P_o stands for the value of P if there were no friction and P is the value with friction.

$$P_o/P = 14.40/23.43 = 0.615 \text{ or } 61.5\%$$

4.7 Pin-in-Slot (Fig. 4.5)

We assume some small clearance between pin and slot, so that contact occurs on one side only. We also recognize that the direction of the friction force of 3 on 2 will be in the direction of the velocity of 3 relative to 2 at the contact point. Any computation routine we set up must determine at what point the contact occurs, and the direction of rubbing at that point. This will require some careful attention to the programming logic.

We first suggest that the situation be represented as shown in Fig. 4.6. Here the friction force at the pin surface has been replaced by an equal force, called $\bar{f}_{32}$, through the pin center plus a couple t_{32}. Unit vector $\bar{u}_t$ is defined to be directed along the common tangent. (Either sense may be chosen.) Unit vector $\bar{u}_n$ is defined as being directed 90 degrees <u>counterclockwise</u> from $\bar{u}_t$. We also define unit vector $\bar{u}_{32}$ to be in the direction of the velocity of 3 relative to 2 at the contact point. With these definitions,

$$\bar{F}_{32}^n = F_{32}^n \, \bar{u}_n$$

$$\bar{f}_{32} = \mu |F_{32}^n| \bar{u}_{32} = \mu |F_{32}^n| D_{32} \, \bar{u}_t$$

where $D_{32} = \bar{u}_{32} \cdot \bar{u}_t$

$$t_{32} = \mu \, R \, F_{32}^n \, D_{32}, \text{ where R is the pin radius.}$$

With this system, Fig. 4.6 will correctly represent any of the four situations pictured in Fig. 4.5. For example, in Fig. 4.5(b) F_{32}^n would be positive, $\bar{u}_{32} = \bar{u}_t$, $D_{32} = +1$, and t_{32} would be positive. In Fig. 4.5(d) F_{32}^n would be negative, $\bar{u}_{32} = \bar{u}_t$, $D_{32} = +1$, and t_{32} would be negative. The representation in Fig. 4.6 is general. The reaction of link 3 on link 2 can be shown in this

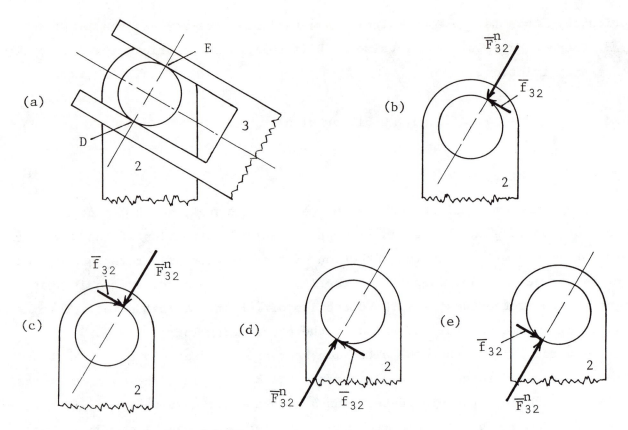

Fig. 4.5   Possible force situations in a pin-in-slot joint.

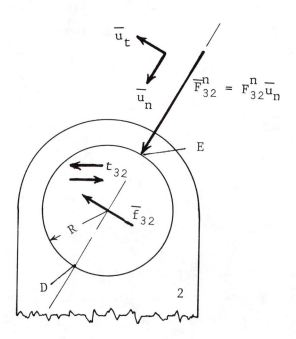

Fig. 4.6   Forces on pin of pin-in-slot joint.   General representation.

way with no prior knowledge of where contact occurs or of the direction of relative motion.

It remains to evaluate $\overline{u}_{32}$ properly. If F_{32}^n is positive, contact is at point E and $\overline{u}_{32}$ must be evaluated at that point. If F_{32}^n is negative, contact is at D and $\overline{u}_{32}$ must be evaluated there. Our program must contain an "IF" statement, depending on the sign of F_{32}^n, and branch to the proper computation of $\overline{u}_{32}$. This will be further illustrated in the example of the next article.

4.8 Example (Fig. 4.7)

$r_2 = 0.50$ in., $\qquad r_1 = 3.00$ in., $\qquad r_4 = 1.25$ in., $\qquad R = 1.00$ in.

$P = 50$ in.-lb$_f$ clockwise, resisting couple on 4.

Q = driving couple on 2, link 2 turning counterclockwise.

$\mu = 0.1$, coefficient of friction between 2 and 3.

Friction at bearings is to be neglected. Inertia is to be neglected.

Q and the internal mechanism reactions are to be determined for the mechanism position where $\Theta_2 = 120$ degrees.

Kinematics

The first step would be to do whatever kinematic analysis will be needed. Here we assume that the basic loop has been solved with the following results at $\Theta_2 = 120$ degrees.

$\Theta_3 = 150.0$ deg., $\qquad r_3 = 3.031$ in.

$h_3 = 0.143$, $\qquad f_3 = 0.429$

It will be necessary to do a little more to establish the direction of relative motion between 2 and 3 at the contact point. We reason as follows.

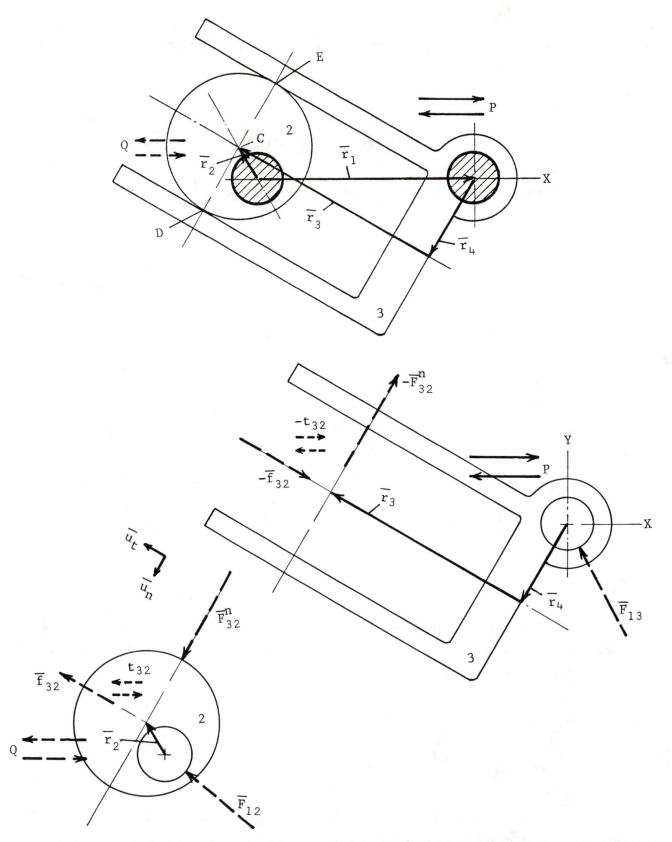

Fig. 4.7   Mechanism for example of Art. 4.8.   Coulomb friction between 2 and 3.

$$\overline{V}_{d_3/d_2} = \overline{V}_{d_3} - \overline{V}_{d_2} = \overline{V}_{c_3} + \overline{V}_{d_3/c_3} - \left(\overline{V}_{c_2} + \overline{V}_{d_2/c_2}\right)$$

$$= \left(\overline{V}_{c_3} - \overline{V}_{c_2}\right) + \overline{V}_{d_3/c_3} - \overline{V}_{d_2/c_2}$$

$$= \overline{V}_{c_3/c_2} + \overline{V}_{d_3/c_3} - \overline{V}_{d_2/c_2}$$

But,

$$\overline{V}_{c_3/c_2} = f_3 \dot{\theta}_2 \overline{u}_t, \text{ where } \overline{u}_t = \text{unit vector in sense of } \overline{r}_3.$$

Also,

$$\overline{V}_{d_3/c_3} = \overline{\omega}_3 \times R\overline{u}_n = -\omega_3 R\overline{u}_t = -h_3 \dot{\theta}_2 R\overline{u}_t$$

$$\overline{V}_{d_2/c_2} = \overline{\omega}_2 \times R\overline{u}_n = -\dot{\theta}_2 R\overline{u}_t$$

Hence,

$$\overline{V}_{d_3/d_2} = (-f_3 - h_3 R + R)\dot{\theta}_2 \overline{u}_t$$

So,

$$\overline{u}_{32} = \left(\overline{V}_{d_3/d_2}\right) \Big/ |V_{d_3/d_2}| = \frac{(-f_3 - h_3 R + R)}{|-f_3 - h_3 R + R|} \frac{\dot{\theta}_2}{|\dot{\theta}_2|} \overline{u}_t \tag{1}$$

In a similar manner we find, if contact is at point E.

$$\overline{u}_{32} = \frac{(-f_3 + h_3 R - R)}{|-f_3 + h_3 R - R|} \frac{\dot{\theta}_2}{|\dot{\theta}_2|} \overline{u}_t \tag{2}$$

For the problem we are working on

$$\overline{u}_{32} = \frac{(-.429 + .143(1) - 1)}{|-.429 + .143(1) - 1|} (1)\overline{u}_t = +\overline{u}_t \text{ at D}$$

$$\overline{u}_{32} = \frac{(-.429 - .143(1) + 1)}{|-.429 - .143(1) - 1|} (1)\overline{u}_t = -\overline{u}_t \text{ at E}$$

4.17

Force and moment equations

Refer to the free-body sketches.

$$F_{32x}^n + f_{32x} + F_{12x} = 0 \tag{3}$$

$$F_{32y}^n + f_{32y} + F_{12y} = 0 \tag{4}$$

$$-F_{32x}^n - f_{32x} + F_{13x} = 0 \tag{5}$$

$$-F_{32y}^n - f_{32y} + F_{13y} = 0 \tag{6}$$

$$r_{2x}F_{32y}^n - r_{2y}F_{32x}^n + r_{2x}f_{32y} - r_{2y}f_{32x} + t_{32} + Q = 0 \tag{7}$$

$$(r_{4x} + r_{3x})(-F_{32y}^n) - (r_{4y} + r_{3y})(-F_{32x}^n) + (r_{4x} + r_{3x})(-f_{32y})$$

$$- (r_{4y} + r_{3y})(-f_{32x}) + (-t_{32}) + P = 0 \tag{8}$$

Numerical substitutions

$$r_{2x} = r_2 \cos \theta_2 = 0.5 \cos 120° = -0.250$$

$$r_{2y} = r_2 \sin \theta_2 = 0.5 \sin 120° = 0.433$$

$$r_{3x} = r_3 \cos \theta_3 = 3.031 \cos 150° = -2.625$$

$$r_{3y} = r_3 \sin \theta_3 = 3.031 \sin 150° = 1.515$$

$$r_{4x} = r_4 \cos \theta_4 = 1.25 \cos 240° = -0.625$$

$$r_{4y} = r_4 \sin \theta_4 = 1.25 \sin 240° = -1.083$$

$$F_{32x}^n = F_{32}^n \cos \theta_n = F_{32}^n \cos 240° = -0.500F_{32}^n$$

4.18

$$F_{32x}^{n} = F_{32}^{n} \sin \Theta_n = F_{32}^{n} \sin 240° = -0.866 F_{32}^{n}$$

$$P = -50 \text{ in.-lb}_f$$

Reduced equations

$$-0.500 F_{32}^{n} + F_{12x} = -f_{32x} \tag{9}$$

$$-0.866 F_{32}^{n} + F_{12y} = -f_{32y} \tag{10}$$

$$0.500 F_{32}^{n} + F_{13x} = f_{32x} \tag{11}$$

$$0.866 F_{32}^{n} + F_{13y} = f_{32y} \tag{12}$$

$$0.433 F_{32}^{n} + Q = 0.250 f_{32y} + 0.433 f_{32x} - t_{32} \tag{13}$$

$$-3.031 F_{32}^{n} - 50 = t_{32} - 3.25 f_{32y} - 0.432 f_{32x} \tag{14}$$

Zero friction solution

Set all friction terms (terms on right side of equations (9)-(14)) equal to zero and solve.

$$F_{32}^{n} = -16.50 \text{ lb}_f$$

$$Q = 7.14 \text{ in.-lb}_f$$

Evaluate friction terms

We first note that, since F_{32}^{n} is negative, the contact point is at "D," so $\overline{u}_{32} = +\overline{u}_t$ and $D_{32} = +1$.

$$\overline{f}_{32} = \mu |F_{32}^{n}| \overline{u}_t = 0.1(16.50)\overline{u}_t = 1.65\overline{u}_t$$

$$\overline{u}_t = \cos 150° \, \overline{i} + \sin 150° \, \overline{j} = -0.866 \, \overline{i} + 0.500 \, \overline{j}$$

$$f_{32x} = -0.866(1.65) = -1.43 \text{ lb}_f$$

$$f_{32y} = 0.500(1.65) = 0.83 \text{ lb}_f$$

$$t_{32} = \mu R F_{32}^n D_{32}$$

$$= 0.1(1)(-16.50)(1) = -1.65 \text{ in.-lb}_f$$

Substitute in equations (13) and (14) and re-solve

$$F_{32}^n = -15.27 \text{ lb}_f$$

$$Q = 7.85 \text{ in.-lb}_f$$

Repeat

$$f_{32} = (.1)(15.27) = 1.53$$

$$f_{32x} = -0.866(1.53) = 1.32$$

$$f_{32y} = 0.500(1.53) = 0.77$$

$$t_{32} = (.1)(1)(-15.27)(1) = -1.53$$

Then,

$$F_{32}^n = -15.33 \text{ lb}_f$$

$$Q = 7.79 \text{ in.-lb}_f$$

This last iteration changed F_{32}^n and Q by less than 1%, so we accept

$$F_{32}^n = -15.3 \text{ lb}_f \quad \text{and} \quad Q = 7.80 \text{ in.-lb}_f$$

4.20

Then, from equations (9)-(12) we calculate the bearing reactions.

$$F_{12x} = -f_{32x} + 0.500F_{32}^{n} = -(-1.3) + 0.500(-15.3) = -6.4 \text{ lb}_f$$

$$F_{12y} = -f_{32y} + 0.866F_{32}^{n} = -0.8 + 0.866(-15.3) = -14.0 \text{ lb}_f$$

$$|F_{12}| = \sqrt{(6.4)^2 + (14.0)^2} = 15.4 \text{ lb}_f$$

$$F_{13x} = f_{32x} - 0.500F_{32}^{n} = 6.4 \text{ lb}_f$$

$$F_{13y} = f_{32y} - 0.866F_{32}^{n} = 14.0 \text{ lb}_f$$

$$|F_{13}| = 15.4 \text{ lb}_f$$

4.9 Straight Slider Joint (Fig. 4.8)

We assume rigid bodies and some clearance in the joints. The various possible situations are represented in sketches (a) - (h) of Fig. 4.8. The block may contact the guide on the top, on the bottom, or be cocked so that one end contacts on top and the other on the bottom. In any of these cases there are two possible directions for the friction forces depending on the direction of motion of 3 relative to 2. The problem is to represent all of these possibilities in some general way so that we can instruct the computer how to deal with whatever arises.

(1) We first take note of the fact that, for those cases in which the block contacts the guide on one side only, the single resultant normal force and corresponding friction force indicated in Fig. 4.8 could be replaced by normal forces and corresponding friction forces at each end of the block. This enables us to treat all eight cases in terms of a pair of normal forces, one at each end of the block. See Fig. 4.9.

(2) We next choose a point around which we plan to take moments (point C in Fig. 4.9), and replace the friction forces at the surface by the equivalent system composed of $\bar{f}_{32}$ through the point plus the couples t_{32} and t'_{32}

4.21

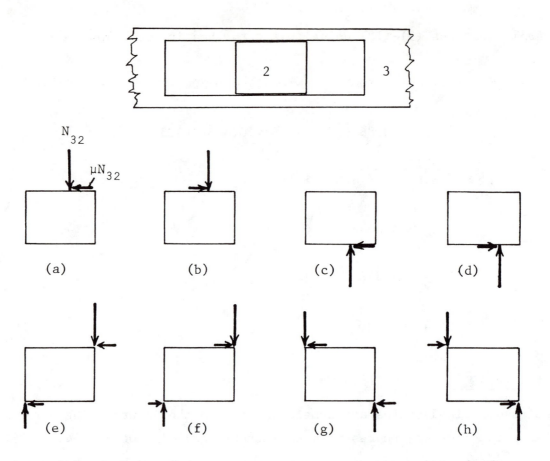

Fig. 4.8   Possible force situations in a straight slider joint.

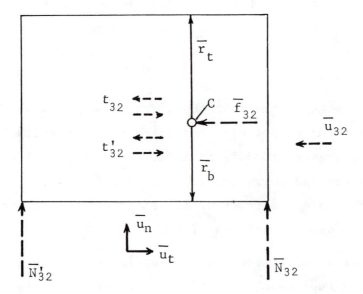

Fig. 4.9   General representation of forces on a straight slider with
Coulomb friction.

$$\overline{f}_{32} = \mu(|N_{32}| + |N'_{32}|)\overline{u}_{32}$$

where $\overline{u}_{32}$ is a unit vector in the direction of motion of 3 relative to 2. This description of $\overline{f}_{32}$ will apply to any of the possible situations.

(3) The next step in the reasoning is to define a unit vector $\overline{u}_t$ parallel to the contact surface and $\overline{u}_n$ normal. $\overline{u}_n$ is defined to be 90 deg. <u>counterclockwise</u> from $\overline{u}_t$. Then,

$$\overline{N}_{32} = N_{32}\overline{u}_n \quad \text{and} \quad \overline{N}'_{32} = N'_{32}\overline{u}_n$$

This will correctly describe the normal forces for any case. N_{32} or N'_{32} positive will mean those forces are acting in the sense of $\overline{u}_n$, negative means the opposite sense. From this we will always know, as the computation progresses, where the block is contacting the guide.

(4) In order to set up expressions for computing the couples t_{32} and t'_{32} we define vectors $\overline{r}_t$ and $\overline{r}_b$ from C to, respectively, the top and bottom surfaces of the block.

$$\overline{r}_t = r_t\overline{u}_n \quad \text{and} \quad \overline{r}_b = r_b\overline{u}_n$$

Then, if we let $D_{32} = \overline{u}_{32} \cdot \overline{u}_t$

$$t_{32} = -\mu r_b N_{32} D_{32} \quad \text{if } N_{32} \text{ is positive}$$

$$= \mu r_t N_{32} D_{32} \quad \text{if } N_{32} \text{ is negative}$$

$$t'_{32} = -\mu r_b N'_{32} D_{32} \quad \text{if } N'_{32} \text{ is positive}$$

$$= \mu r_t N'_{32} D_{32} \quad \text{if } N'_{32} \text{ is negative}$$

These "if" statements must be incorporated into our computing routine at the point where we need to compute t_{32} and t'_{32}.

To illustrate the actual operating procedure for implementing the above suggestions we will solve a simple example.

4.10 Example, Straight Slider Joint (Fig. 4.10)

The block is moving to the left.

Force and moment equations

$$P_x + Q_x + N_{14x} + N'_{14x} + f_{14x} = 0 \qquad (1)$$

$$P_y + Q_y + N_{14y} + N'_{14y} + f_{14y} = 0 \qquad (2)$$

$$f_x N_{14y} - r_y N_{14x} + r'_x N'_{14y} - r'_y N_{14x} + r_{px} P_y - r_{py} P_x$$

$$+ t_{14} + t'_{14} = 0 \qquad (3)$$

Values to be inserted

$$P_y = 0, \qquad\qquad P_x = P$$

$$Q_x = Q \cos \Theta_q = 100 \cos 330° = 86.6 \text{ lb}_f$$

$$Q_y = Q \sin \Theta_q = 100 \sin 330° = -50.0$$

$$r_x = 2.00 \text{ in.} \qquad\qquad r_y = 0$$

$$r'_x = -2.00 \text{ in.} \qquad\qquad r'_y = 0$$

$$r_{px} = 0 \qquad\qquad r_{py} = 1.00 \text{ in.}$$

$$N_{14x} = 0 \qquad\qquad N_{14y} = N_{14}$$

$$N'_{14x} = 0 \qquad\qquad N'_{14y} = N'_{14}$$

$$f_{14y} = 0 \qquad\qquad f_{14x} = f_{14}$$

4.24

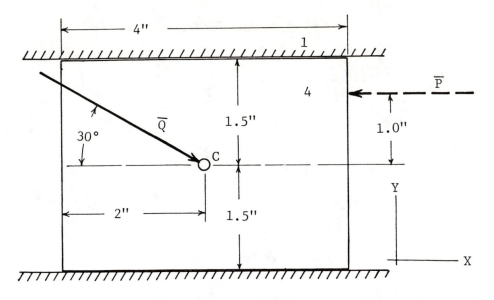

(a)   Problem description.

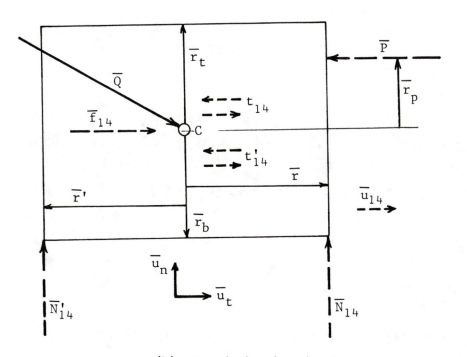

(b)   Freebody sketch.

Fig. 4.10   Problem of article 4.10.

4.25

Reduced equations

$$P + 86.60 = -f_{14} \tag{4}$$

$$-50 + N_{14} + N'_{14} = 0 \tag{5}$$

$$2.00\,N_{14} - 2.00\,N'_{14} - 1.00\,P = -t_{14} - t'_{14} \tag{6}$$

Zero-friction solution

Set $f_{14} = t_{14} = t'_{14} = 0$ and solve above equations.

$$P = -86.60$$

$$N_{14} = 3.35 \qquad\qquad N'_{14} = 46.65$$

Evaluate friction terms

$$D_{14} = \overline{u}_{14} \cdot \overline{u}_t = +1$$

$$f_{14} = \mu(|N_{14}| + |N'_{14}|)D_{14} = 0.2(3.35 + 46.65)(+1) = 10.00$$

$$\overline{r}_t = r_t\overline{u}_n = 2.00\overline{u}_n, \qquad r_t = 2.00$$

$$\overline{r}_b = r_b\overline{u}_n = -2.00\overline{u}_n, \qquad r_b = -2.00$$

Since N_{14} is positive

$$t_{14} = -r_b\mu N_{14}D_{14} = -(-2.00)(.2)(3.35)(+1) = +1.34 \text{ in.-lb}_f$$

Since N'_{14} is positive

$$t'_{14} = -r_b\mu N'_{14}D_{14} = -(-2.00)(.2)(46.65)(+1) = +18.60 \text{ in.-lb}_f$$

4.26

$$P = -96.60 \text{ lb}_f \qquad N'_{14} = 52.90 \text{ lb}_f$$

$$N_{14} = -2.90 \text{ lb}_f$$

Re-evaluate friction terms

$$f_{14} = 0.2(2.90 + 52.90)(+1) = 11.16$$

Since N_{14} is now <u>negative</u>.

$$t_{14} = r_t \mu N_{14} D_{14} = (2.00)(.2)(-2.90)(+1) = -1.16 \text{ in.-lb}_f$$

Since N'_{14} is positive

$$t'_{14} = -r_b \mu N'_{14} D_{14} = -(-2.00)(.2)(52.90)(+1) = 21.16 \text{ in.-lb}_f$$

Continue, re-solving the force and moment equations and re-evaluating the friction terms until the results of successive iterations show no significant change. The table of Fig. 4.11(a) shows the results for this problem through five solutions of the equations.

Observe how the process of computation takes care of the question of where the block contacts the guide. At the end of step (1) both N_{14} and N'_{14} were positive, indicating contact on the bottom only. At the end of step (2) N_{14} was negative, indicating counterclockwise cocking of the block relative to the guide. This situation was maintained in the final solution.

4.11 Including Inertia Effects

All the examples so far presented were <u>static</u> force analyses. If inertia effects are included, the effect on the problem is to add more known terms into the force and moment equations. The procedure for computation would not be changed except for the addition of the necessary kinematic computations to establish values of the accelerations in order to compute "$m\overline{A}_g$" and "$I\alpha$" terms.

	(1)	(2)	(3)	(4)	(5)
f_{14}	0	10.00	11.16	11.78	11.84
t_{14}	0	1.34	-1.16	-1.78	-1.84
t'_{14}	0	18.60	21.16	21.76	21.84
P	-86.60	-96.60	-97.76	-98.38	-98.44
N_{14}	3.35	-2.90	-4.44	-4.59	-4.61
N'_{14}	46.65	52.90	54.44	54.59	54.61

(a) Progress of the iterative computations.

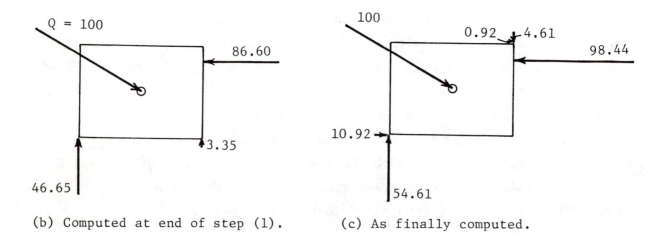

(b) Computed at end of step (1).   (c) As finally computed.

Fig. 4.11   Results of computations for problem of article 4.10.

4.28

E4.1 For the mechanism studied in article 4.5 assume Q to be driving (link 2 turning clockwise) and P to be the resisting load. Take all other data to be the same as before. Do the iterative computation and present the results in a table like Fig. 4.4. Also determine the instantaneous efficiency of the mechanism.

E4.2 Show how to program the static force analysis of this mechanism with
      Coulomb friction at all joints. Complete the freebody sketches. Care-
      fully define all symbols used. Write equations to be solved. Explain
      any special logic needed in the program. Take Q to be an unknown driving
      couple applied to link 2 and P to be a known resisting force applied to
      link 3.

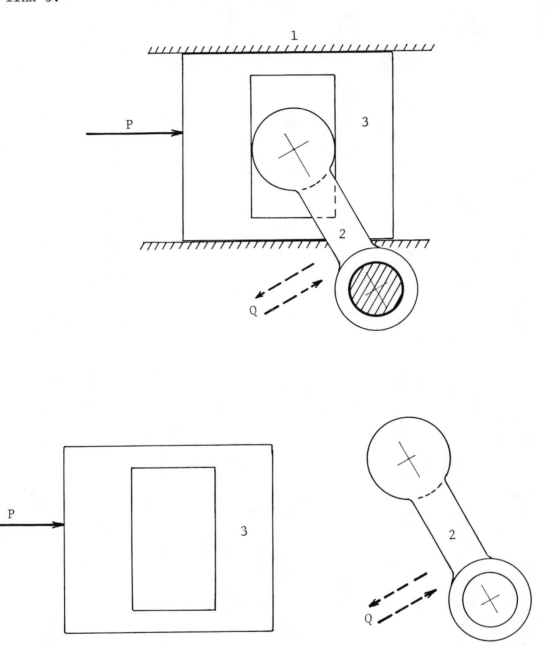

E4.3 Explain how you would suggest dealing with this situation in a force
analysis including Coulomb friction. Define all symbols used. Explain
any special logic that would be needed in a computer program.

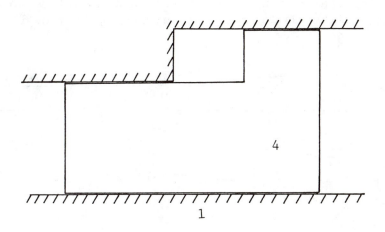

E4.4 Q is a known driving couple applied to link 2. P is an unknown resisting force applied to link 4. Coulomb friction is to be considered at
all joints. Inertia is to be neglected.

Complete the freebody sketches. Write all equations to be solved and
explain the logic of a computer program to determine P and all internal
forces.

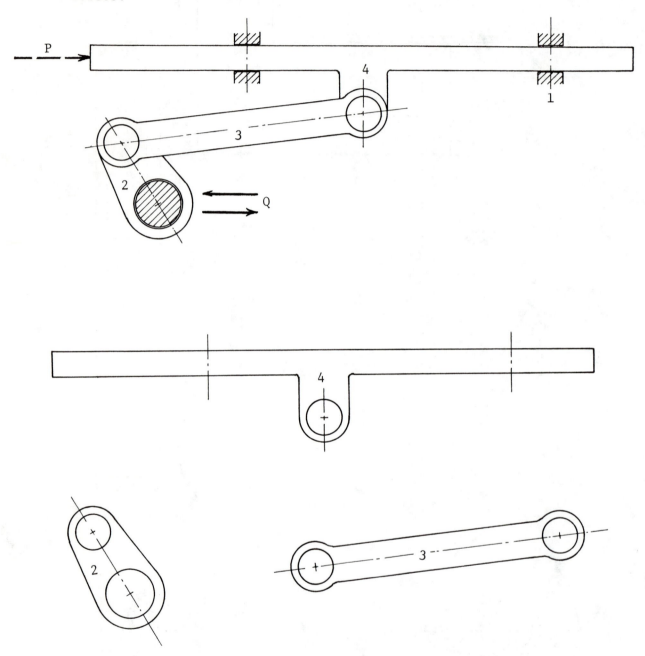

E4.5 P is a known resisting force applied to link 4. Q is an unknown driving
      couple applied to crank 2. Coulomb friction is to be considered at all
      joints. Inertia is to be neglected.

      Complete the freebody sketches and write all force and moment equations
      for a static force analysis. Explain the logic of a computer program
      for the computations.

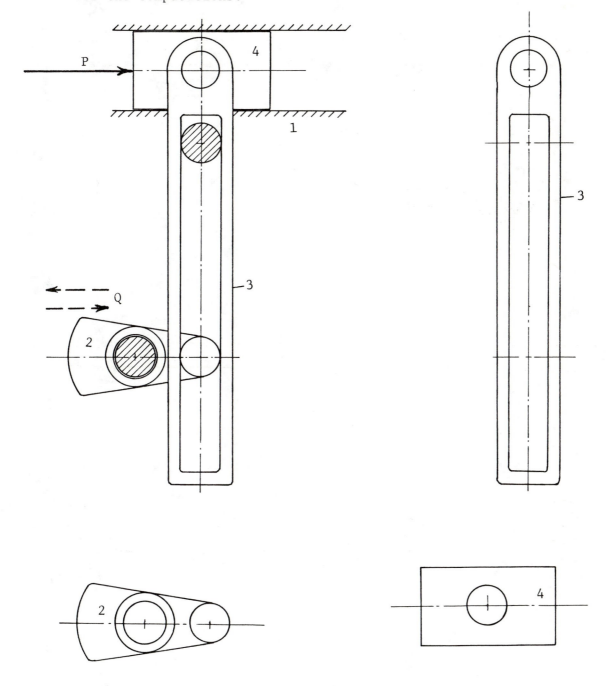

4.33

CHAPTER 5

POWER EQUATION. EQUATION OF MOTION

5.1 Power Equation

An energy balance for a machine can be written as follows.

$$W = \Delta T + \Delta U + W_f \tag{1}$$

where,

   W = net work input to the machine

     = (work in) - (work out).

   ΔT = change in kinetic energy of the moving parts.

   ΔU = change in potential energy stored in the machine.

   W_f = energy dissipated through friction.

In terms of <u>time</u> <u>rates</u> we have the <u>power</u> <u>equation</u>.

$$P = dT/dt + dU/dt + P_f \tag{2}$$

where,

   dT/dt = rate of change of stored kinetic energy.

   dU/dt = rate of change of stored potential energy.

   P_f = rate of dissipation of energy.

In high-speed machines the kinetic energy is usually the predominant term on the right-hand side of the power equation. In slow moving, heavy machinery potential energy due to elevation of the moving masses may be significant. Potential energy stored in elastic members (springs) may be important in some devices.

Energy dissipation takes place in the bearings and joints of a machine as the parts rub on each other. The effect of this is to raise the temperature of the surrounding parts and to cause heat flow from the machine to the environment. This energy is "lost" in the sense that it subtracts from the possible useful work output of the machine. Energy dissipating elements may be deliberately built into a machine for various purposes. Examples are brakes, clutches, and "shock absorbers."

In the next few articles we look in more detail at descriptions of various terms in the power equation.

5.2 Kinetic Energy (Fig. 5.1)

The kinetic energy, T, of a rigid body in plane motion is

$$T = \tfrac{1}{2}mV^2 + \tfrac{1}{2}I\omega^2 \tag{1}$$

where,

   m = mass of body

   V = velocity of the center of mass

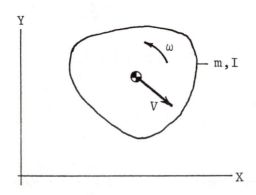

Fig. 5.1 Kinetic energy, T, of a body in plane motion.

ω = angular velocity of the body

I = moment of inertia of the body about the axis through the center
of mass perpendicular to the plane of motion.

Let X,Y be the coordinates of the body center of mass, Θ measure the angular position of the body, and S_i be the input position variable for the mechanism to which the body belongs. Then the following kinematic coefficients can be defined.

$$f_x = dX/dS_i, \quad f_y = dY/dS_i \quad \text{and} \quad h = d\Theta/dS_i.$$

In terms of these coefficients, and the input velocity $\dot{S}_i$, the kinetic energy is

$$T = [\tfrac{1}{2}m(f_x^2 + f_y^2) + \tfrac{1}{2}Ih^2]\dot{S}_i^2 \tag{2}$$

where,

$\dot{S}_i$ is the input velocity.

The time rate of change of this kinetic energy would be

$$dT/dt = [m(f_x^2 + f_y^2) + Ih^2]\dot{S}_i\ddot{S}_i + [m(f_x f_x' + f_y f_y') + Ihh']\dot{S}_i^3$$

$$= A\dot{S}_i\ddot{S}_i + B\dot{S}_i^3 \tag{3}$$

where A and B are functions of the input variable, S_i.

Note: $d(f_x^2)/dt = d(f_x^2)/dS_i)(dS_i/dt)$

$$= 2f_x(df_x/dS_i)(dS_i/dt) = 2f_x f_x'\dot{S}_i$$

For a machine the time rate of change of kinetic energy would be

$$dT/dt = (\Sigma A)\dot{S}_i\ddot{S}_i + (\Sigma B)\dot{S}_i^3 \tag{4}$$

where ΣA and ΣB are the sums of A and B for all moving parts of the machine.

5.3 Equivalent Mass or Moment of Inertia

Consider the term called "(ΣA)" in equation (4) of the previous article.

$$(\Sigma A) = \Sigma[m(f_x^{\,2} + f_y^{\,2}) + Ih^2]$$

This term has the units of mass if S_i is a length, r_i. It has the units of moment of inertia if S_i is an angle, Θ_i. This is sometimes called the "equivalent" mass (m_e) or "equivalent" moment of inertia (I_e) of the machine. It is the mass or moment of inertia which, if attached to the input link, would have the same kinetic energy as the total machine.

$$T = \tfrac{1}{2}(\Sigma A)\ \dot{S}_i^{\,2} = \tfrac{1}{2}m_e\dot{r}_i^{\,2} \quad \text{or} \quad \tfrac{1}{2}I_e\dot{\Theta}_i^{\,2}$$

The term called "(ΣB)" is

$$(\Sigma B) = \Sigma[m(f_x f_x' + f_y f_y') + Ihh'] = \frac{1}{2}\frac{d(\Sigma A)}{dS_i}$$

$$= \frac{1}{2}\frac{dm_e}{dr_i}, \quad \text{or} \quad \frac{1}{2}\frac{dI_e}{d\Theta_i}$$

$$dT/dt = I_e\dot{\Theta}_i\ddot{\Theta}_i + \frac{1}{2}\frac{dI_e}{d\Theta_i}\dot{\Theta}_i^{\,3}, \quad \text{for angle input } \Theta_i$$

$$dT/dt = m_e\dot{r}_i\ddot{r}_i + \frac{1}{2}\frac{dm_e}{dr_i}\dot{r}_i^{\,3}, \quad \text{for linear input } r_i$$

5.4 Power Equation. Example (Fig. 5.2)

The mechanism shown is the same one used in article 3.5 (Fig. 3.6) for an example force analysis. From the work already done there we have, or can very easily obtain, the following results of the kinematic analysis.

$$h_3 = 0.2857 \quad (= h_4)$$

$$h_3' = 0.1060 \quad (= h_4')$$

5.4

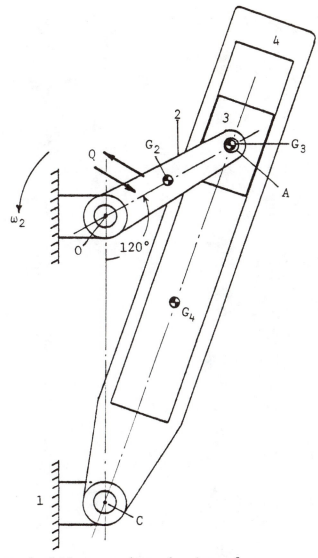

Fig. 5.2 Inverted slider-crank mechanism, for power equation example.

OA = 6 in., OC = 12 in., OG_2 = 3 in., CG_4 = 9 in.

m_2 = 3 lb_m, m_3 = 2 lb_m, m_4 = 8 lb_m, I_2 = 12 $lb_m in^2$

I_3 = 0.5 $lb_m in^2$, I_4 = 200 $lb_m in^2$

Q = driving couple.

$$f_{x2}(= dX_{g2}/d\Theta_2) = -1.500$$

$$f_{y2}(= dY_{g2}/d\Theta_2) = 2.598$$

$$f_{x3}(= dX_{g3}/d\Theta_2) = -3.000$$

$$f_{y3}(= dY_{g3}/d\Theta_2) = 5.196$$

$$f_{x4}(= dX_{g4}/d\Theta_2) = -2.430$$

$$f_{y4}(= dY_{g4}/d\Theta_2) = 0.842$$

$$f'_{x2}(d^2X_{g2}/d\Theta_2^2) = 2.598$$

$$f'_{y2}(= d^2Y_{g2}/d\Theta_2^2) = -1.500$$

$$f'_{x3}(= d^2X_{g3}/d\Theta_2^2) = -5.196$$

$$f'_{y3}(= d^2Y_{g3}/d\Theta_2^2) = -3.000$$

$$f'_{x4}(= d^2X_{g4}/d\Theta_2^2) = -1.142$$

$$f'_{y4}(= d^2Y_{g4}/d\Theta_2^2) = -0.382$$

Then we can compute the "A" and "B" terms in the power equation.

$$A_2 = m_2(f_{x2}^{\ 2} + f_{y2}^{\ 2}) + I_2h_2^2$$

$$= 3[(-1.500)^2 + (2.598)^2] + 12(1)^2 = 39.00 \text{ lb}_m\text{ in.}^2$$

$$A_3 = m_3(f_{x3}^{\ 2} + f_{y3}^{\ 2}) + I_3h_3^2$$

$$= 2[(-3.000)^2 + (5.196)^2] + 0.5(0.2857)^2 = 72.04 \text{ lb}_m\text{ in.}^2$$

$$A_4 = m_4(f_{x4}{}^2 + f_{y4}{}^2) + I_4 h_4{}^2$$

$$= 8[(-2.430)^2 + (0.842)^2] + 200(0.2857)^2 = 69.24 \text{ lb}_m \text{in.}^2$$

$$B_2 = m_2(f_{x2}f'_{x2} + f_{y2}f'_{y2}) + I_2 h_2 h'_2$$

$$= 3[(-1.500)(-2.598) + (2.598)(-1.500)] + 12(1)(0) = 0$$

$$B_3 = m_3(f_{x3}f'_{x3} + f_{y3}f'_{y3}) + I_3 h_3 h'_3$$

$$= 2[(-3.000)(-5.196) + (5.196)(-3.000)] + 0.5(0.2857)(0.1060)$$

$$= .02 \text{ lb}_m \text{in.}^2$$

$$B_4 = m_4(f_{x4}f'_{x4} + f_{y4}f'_{y4}) + I_4 h_4 h'_4$$

$$= 8[(-2.430)(-1.142) + (0.842)(-0.382)] + 200(0.2857)(0.1060)$$

$$= 25.68 \text{ lb}_m \text{in.}^2$$

$$(\Sigma A) = A_2 + A_3 + A_4 = 180.28 \text{ lb}_m \text{in.}^2, \quad \text{or } 3.888 \times 10^{-2} \text{ slug-ft}^2$$

$$(\Sigma B) = B_2 + B_3 + B_4 = 25.70 \text{ lb}_m \text{in.}^2, \quad \text{or } 0.554 \times 10^{-2} \text{ slug-ft}^2$$

Hence the time rate of change of kinetic energy for this machine, in this position of the driving crank 2, is

$$dT/dt = (\Sigma A)\, \dot\theta_2 \ddot\theta_2 + (\Sigma B)\, \dot\theta_2{}^3$$

$$(3.888 \times 10^{-2})\dot\theta_2 \ddot\theta_2 + (0.554 \times 10^{-2})\dot\theta_2{}^3 \text{ lb}_f\text{-ft/sec}$$

$$(\dot\theta_2 \text{ in rad/sec}, \quad \ddot\theta_2 \text{ in rad/sec}^2)$$

In the problem of article 3.5 it was specified the only external loading was couple Q applied to link 2. Therefore the net power input to the machine was $Q\dot{\theta}_2$ and the power equation would be

$$Q\dot{\theta}_2 = (3.888 \times 10^{-2})\dot{\theta}_2\ddot{\theta}_2 + (0.554 \times 10^{-2})\dot{\theta}_2^3$$

If we divide by $\dot{\theta}_2$, we have the <u>equation</u> <u>of</u> <u>motion</u>.

$$Q = (3.888 \times 10^{-2})\ddot{\theta}_2 + (0.554 \times 10^{-2})\dot{\theta}_2^2$$

In the same problem it was also specified that $\dot{\theta}_2$ = 40 rad/sec and $\ddot{\theta}_2$ = 0, hence

$$Q = (0.554 \times 10^{-2})(40)^2 = 8.864 \ lb_f\text{-ft}, \quad or \quad 106.4 \ lb_f\text{-in.}$$

This is the same result obtained from the force analysis.

The same equation of motion holds regardless of the specifications on Q, $\dot{\theta}_2$ and $\ddot{\theta}_2$. For example, suppose we specify Q = 20 lb_f-ft and $\dot{\theta}_2$ = 80 rad/sec and ask what would be the resulting angular acceleration $\ddot{\theta}_2$ of the driving crank.

$$\ddot{\theta}_2 = \frac{Q - (0.554 \times 10^{-2})\dot{\theta}_2^2}{3.888 \times 10^{-2}}$$

$$= \frac{20 - (0.554 \times 10^{-2})(80)^2}{3.888 \times 10^{-2}} = -398 \ rad/sec^2$$

The crank would be <u>slowing</u> <u>down</u> at the rate of 398 rad/sec^2.

If we were to add other external couples or forces acting on the machine this would change the description of the net power input but would not affect the terms on the right-hand side of the power equation. Suppose that we specify unknown couple Q acting on crank 2, a couple P = 50 lb_f-ft clockwise on link 4, $\dot{\theta}_2$ = 40 rad/sec, counterclockwise and θ_2 = 200 rad/sec^2 counterclockwise. The net power input to the machine is then

$$Q\dot{\theta}_2 + P\dot{\theta}_4 = (Q + Ph_4)\dot{\theta}_2$$

The power equation is

$$(Q + Ph_4)\dot{\theta}_2 = (\Sigma A)\ddot{\theta}_2\dot{\theta}_2 + (\Sigma B)\dot{\theta}_2^3$$

After dividing through by $\dot{\theta}_2$ we have the equation of motion.

$$Q + Ph_4 = (\Sigma A)\ddot{\theta}_2 + (\Sigma B)\dot{\theta}_2^2$$

Substituting the numbers for this problem and solving for Q we have

$$Q = (\Sigma A)\ddot{\theta}_2 + (\Sigma B)\dot{\theta}_2^2 - Ph_4$$

$$= (3.888 \times 10^{-2})(200) + (0.554 \times 10^{-2})(40)^2 - (-50)(0.2857)$$

$$= 30.2 \; lb_f\text{-ft}$$

5.5 Potential Energy, Elevation (Fig. 5.3)

The potential energy due to elevation, H, of a rigid body above some arbitrary reference level is

$$U_e = mgH \tag{1}$$

where,

m = mass of the body

g = gravitational acceleration magnitude

H = vertical distance from reference level to the body center of mass.

The time rate of change of this potential energy is

$$dU_e/dt = mg(dH/dt) = mgV_e \tag{2}$$

where,

V_e = the vertical component of the velocity of the center of mass.

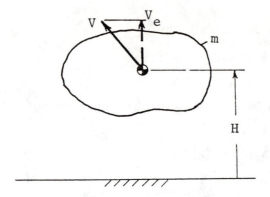

Fig. 5.3 Reference sketch. Potential energy of elevation.

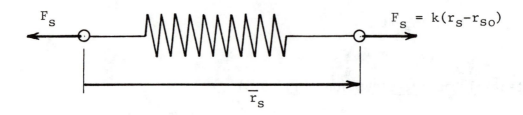

$$F_s = k(r_s - r_{so})$$

Fig. 5.4 Reference sketch. Potential energy of linear spring.

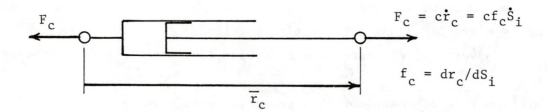

$$F_c = c\dot{r}_c = cf_c\dot{S}_i$$

$$f_c = dr_c/dS_i$$

Fig. 5.5 Reference sketch. Viscous damper.

$$V_e = (dH/dS_i)(dS_i/dt) = f_e \dot{S}_i$$

so,

$$dU_e/dt = (mgf_e)\dot{S}_i \qquad (3)$$

5.6 Potential Energy, Linear Spring (Fig. 5.4)

The potential energy of a linear spring is

$$U_s = \tfrac{1}{2}k(\Delta\ell)^2 \qquad (1)$$

where,

k = spring constant, force/deflection

and

$\Delta\ell$ = length of spring – unloaded length of spring

If $\overline{r}_s$ is the vector running from one end of the spring to the other,

$$U_s = \tfrac{1}{2}k(r_s - r_{so})^2 \qquad (2)$$

where,

r_{so} is the length of that vector when the spring is unloaded.

The time rate of change of energy is

$$dU_s/dt = k(r_s - r_{so})\dot{r}_s = k(r_s - r_{so})f_s \dot{S}_i \qquad (3)$$

5.7 Energy Dissipated, Viscous Damper (Fig. 5.5)

The magnitude of the force required to move the elements of a viscous damper relative to each other is

$$F_c = cV_{rel} \qquad (1)$$

5.11

where,

c = damping coefficient, force/velocity

and

V_{rel} = magnitude of the relative velocity between damper elements.

The work done to move the damper elements a small distance, $d\ell$, relative to each other is

$$dW = cV_{rel}d\ell \qquad (2)$$

Hence, the <u>rate</u> of doing work on the damper, and consequently the rate at which energy is dissipated in the damper, is

$$dW/dt = cV_{rel}d\ell/dt \qquad (3)$$

In terms of symbols used in Fig. 5.5

$$dW/dt = c\dot{r}_c^2 = (cf_c^2)\dot{s}_i^2 \qquad (4)$$

5.8 Energy Dissipation - Coulomb Friction

In order to compute the rate of energy dissipation through Coulomb friction, we must know the friction force and the rubbing velocity. This means that a force analysis, in addition to the kinematic analysis, must be done in order to evaluate this dissipation term in the power equation. Figure 5.6 reproduces sketches from chapter 4 showing the suggested treatment of Coulomb friction in pin, pin-in-slot, and straight slider joints. The rate of energy dissipation, P_f, in these cases can be described as follows:

	P_f	
Pin-joint	$t_{32}\omega_{32}$	
Pin-in-slot	$f_{32}V_{d_3/d_2}$	(contact at D)
	$f_{32}V_{e_3/e_2}$	(contact at E)
or,	$f_{32}V_{c_3/c_2} + t_{32}\omega_{32}$	(either case)
Straight-slider	$f_{32}V_{3/2}$	

5.12

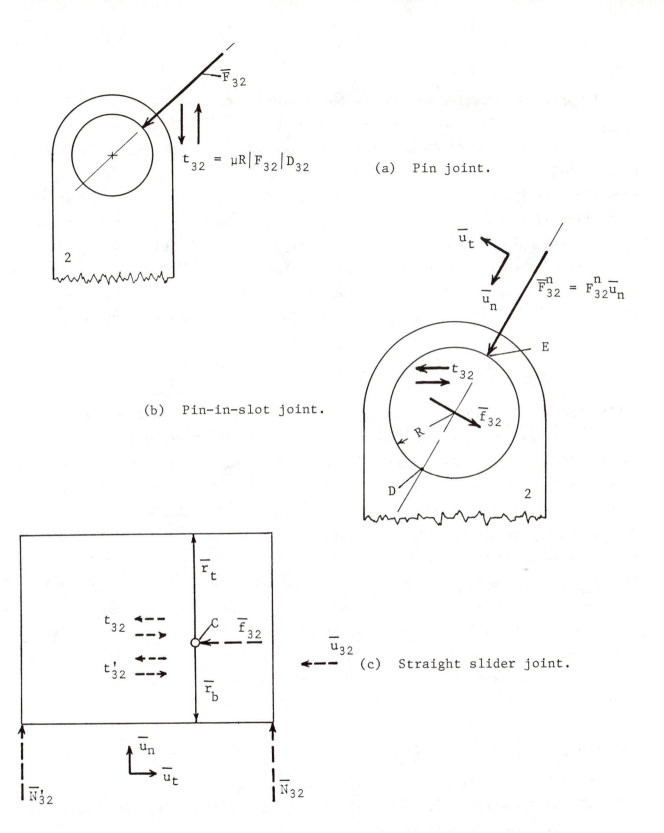

$t_{32} = \mu R |F_{32}| D_{32}$ (a) Pin joint.

(b) Pin-in-slot joint.

$\overline{F}_{32}^n = F_{32}^n \overline{u}_n$

(c) Straight slider joint.

Fig. 5.6 Coulomb friction in different joints.

5.9 Equation of Motion

In previous articles we have rather casually referred to the equation of motion as the equation obtained after dividing through the power equation by the input velocity. This is an equation involving the input variable, its first two time derivatives, and the external forces or couples. It is a differential equation. If the state of motion (input position, velocity and acceleration) is completely known, then this can be used to calculate the relationship between the external forces or couples, which is what we did in article 5.4. In the latter part of that article we assumed, for the mechanism of Fig. 5.2, that the state of motion was known and that there was a known couple, P, acting on link 4 plus an unknown couple, Q, acting on input link 2. The equation of motion was

$$Q = (\Sigma A)\ddot{\theta}_2 + (\Sigma B)\dot{\theta}_2^2 - Ph_4$$

Everything on the right side could be calculated, hence Q could be evaluated.

Now let us consider the same mechanism but a different, and perhaps more interesting, problem. Assume that Q and P are given and that we know the state of motion at time t = 0. How can we calculate the state of motion at any later time t? The answer, of course, is to solve the differential equation to determine θ_2 as a function of time. However, this is not simply a matter of referring to a text on differential equations for an answer. The equation is not of a nature for which we are likely to find a closed-form solution. It is non-linear and the "coefficients" (ΣA) and (ΣB) are not constants, but are functions of θ_2. Hence the most feasible way to solve the problem is a numerical, step-by-step procedure. Several such procedures have been derived and are available as standard computer sub-routines. Check your local computer center for documentation on what is available.

5.10 Example. Equation of Motion

To emphasize a bit more what is involved in setting up and preparing to solve the equation of motion for a mechanism, we continue with the problem of Fig. 5.2, with this additional information.

(1) The machine is driven by an electric motor, coupled to the shaft of link 2, whose torque-speed characteristic is

$$Q = 30[1 - (\dot{\theta}_2/125.7)^3] \text{ in.-lb}_f$$

$$(\dot{\theta}_2 \text{ in rad/sec})$$

(2) Motor rotation is counterclockwise

(3) The load torque, P, is variable and proportional to the angular velocity of link 4.

$$P = -40\dot{\theta}_4 = -40 \; h_4\dot{\theta}_2$$

(4) The motor will be started at t = 0 with the machine in position $\theta_2 = 0$. Hence initial conditions on the differential equation are

$$\theta_2 = 0, \quad \dot{\theta}_2 = 0 \quad \text{at} \quad t = 0$$

The problem we are interested in is the determination of θ_2 and $\dot{\theta}_2$ as functions of time from t = 0 until the time at which approximate steady-state conditions are reached.

We first substitute the given information concerning Q and P into the equation of motion as written in the previous article.

$$Q = (\Sigma A)\ddot{\theta}_2 + (\Sigma B)\dot{\theta}_2^2 - Ph_4 \tag{1}$$

$$30[1 - (\dot{\theta}_2/125.7)^3] = (\Sigma A)\ddot{\theta}_2 + (\Sigma B)\dot{\theta}_2^2 + 40h_4^2\dot{\theta}_2 \tag{2}$$

Solving for $\ddot{\theta}_2$, we have

$$\ddot{\theta}_2 = [1/(\Sigma A)][30 - 30(\dot{\theta}_2/125.7)^3 - (\Sigma B)\dot{\theta}_2^2 - 40h_4^2\dot{\theta}_2] \tag{3}$$

A rough outline of the computing program would be as follows:

1. Enter all geometric and physical data and the initial conditions. Also state the time step size, Δt, to be used.

2. Call the sub-program to be used in solving the differential equation.

3. Within this sub-program place all the computations needed to
    establish numerical values for (ΣA), (ΣB) and h_4. Then write
    the differential equation. The program will compute and return
    the values of Θ_2 and $\dot{\Theta}_2$ at the end of the first time interval.

4. Increment t and recall the sub-program. Repeat until some
    previously established limit on t, or Θ_2, or $\dot{\Theta}_2$ is reached.

The above is a very broad outline. Your instructor will probably discuss
this in more detail. Also be sure to follow very carefully whatever documentation you are using on the differential equation solving program employed. Here
we will make only a few more remarks and suggestions.

(1) Be very sure that a <u>consistent set of units</u> is used throughout.
     This is perhaps the most common mistake made.

(2) Since the machine starts at $\dot{\Theta}_2 = 0$ and picks up speed, the time
     interval should be varied to maintain reasonable incremental values
     of Θ_2. At the appropriate point in your program place an "IF" state-
     ment of the following nature.

$$\text{If } \left| \Theta_{2\,\text{current}} - \Theta_{2\,\text{previous}} \right| \text{ is greater than 10 deg.,}$$

     then make $\Delta t = \Delta t/2$.

(3) The choice of Δt to start with could be a random, or "hunch" value,
     but it would be better to establish something on a more rational
     basis. If we look at the differential equation at t = 0, we see

$$\ddot{\Theta}_2 = 30/(\Sigma A)$$

     From a skeleton scale drawing of the mechanism in the initial
     position (Fig. 5.7) we can quickly establish approximate numerical
     values for the kinematic coefficients needed to calculate (ΣA).

$$h_2 = 1$$

$$h_3 = h_4 = (I_{12}I_{24})/(I_{14}I_{24}) = (3.0)/(15.0) = 0.20$$

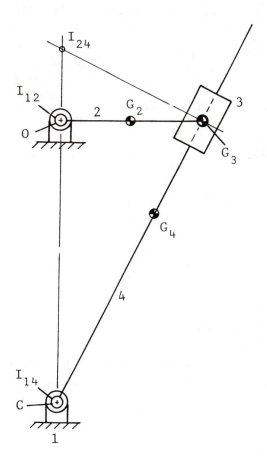

Fig. 5.7 Skeleton drawing to evaluate constants in equation of motion
in initial position of mechanism.

$$f_{x2}^2 + f_{y2}^2 = (OG_2)^2 = (3)^2 = 9 \text{ in.}^2$$

$$f_{x3}^2 + f_{y3}^2 = (OG_3)^2 = (6)^2 = 36 \text{ in.}^2$$

$$f_{x4}^2 + f_{y4}^2 = (CG_4)^2 h_4^2 = (9)^2(.20)^2 = 3.24 \text{ in.}^2$$

$$A_2 = m_2(f_{x2}^2 + f_{y2}^2) + I_2 h_2^2$$

$$= 3(9) + 12(1)^2 = 39 \text{ lb}_m \text{ in.}^2$$

$$A_3 = 2(36) + 0.5(.20)^2 \approx 72 \text{ lb}_m \text{ in.}^2$$

$$A_4 = 8(3.24) + 200(.20)^2 \approx 34 \text{ lb}_m \text{ in.}^2$$

$$(\Sigma A) \approx 145 \text{ lb}_m \text{ in.}^2 , \quad \text{or} \quad \approx 4.5 \text{ slug-in.}^2$$

So:

$$\ddot{\Theta}_2 \approx (30)(12)/4.5 \approx 80 \text{ rad/sec}^2$$

If we assume that, over the time interval Δt, the acceleration is nearly constant, then

$$\Delta\Theta_2 \approx \tfrac{1}{2}\ddot{\Theta}_2(\Delta t)^2$$

If we require the first $\Delta\Theta_2$ to be approximately 10 deg., then

$$\Delta t = \sqrt{2(10)(\pi/180)/80} \approx 0.066 \text{ sec.}$$

So a reasonable starting value for Δt would be 0.06 or 0.05 seconds.

E5.1 In the ordinary gear train shown let the letters a,b....g stand for the
      tooth numbers of the indicated gears. In terms of these tooth numbers,
      and appropriate symbols for other geometric and physical constants as
      needed, determine the equivalent moment of inertia of the train. Take
      shaft 2 to be the input element.

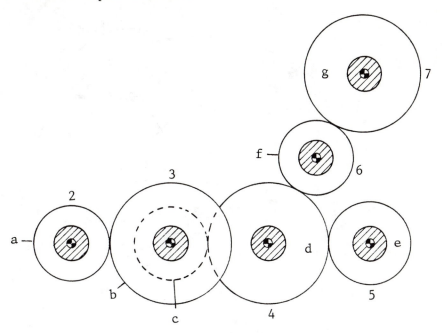

E5.2 In the epicyclic train shown let the letters a,b,c,d stand for the tooth
      numbers of the indicated gears. In terms of these tooth numbers, and
      appropriate symbols for other geometric and physical constants as needed,
      determine the equivalent moment of inertia of the train. Take the epi-
      cyclic arm, 2, to be the input element.

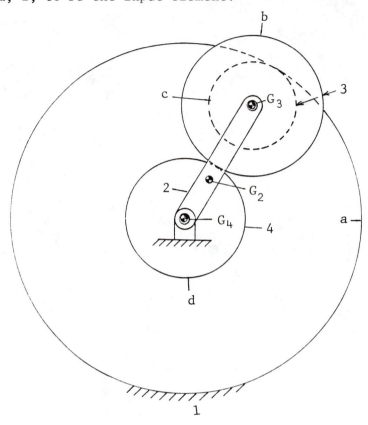

E5.3 Refer to the problem of article 4.5. Using the results of the force
analysis carried out there, compute the power dissipated and compare
with the net power input. Does this provide a check on the correctness
of the force analysis computations?

E5.4 Assume $\overline{P}$ to be given and Q (applied to crank 2) to be unknown. All geo-
      metric and physical constants are known, as well as the position, angular
      velocity and angular acceleration of crank 2.

      Outline the computations needed to establish Q via the power equation.

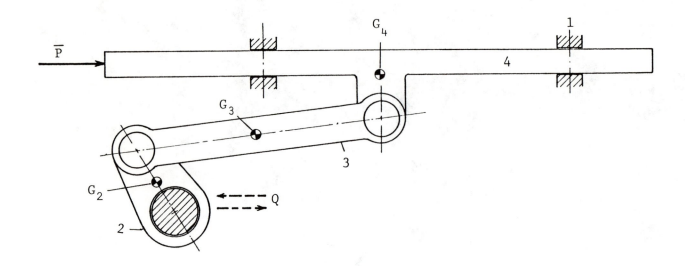

E5.5 Assume that both $\overline{P}$ and Q are given. All geometric and physical constants are known. The initial position and angular velocity of crank 2 is given.

Program the computations needed to solve the equation of motion.

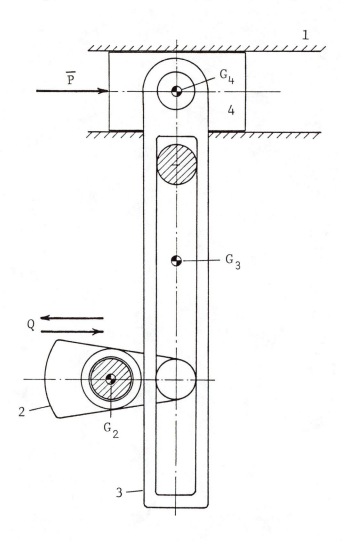

E5.6 Assume link 2 to be the input and work up a program to solve the differential equation of motion. Crank 2 is driven counter-clockwise by a
linear spring having a spring constant of k in-lb_f/rad, and an initial
torque (at $\Theta_2 = 0$) of Q_0 in-lb_f.

At t = 0, $\Theta_2 = 0$ and $\dot{\Theta}_2 = 0$. The objective is to determine Θ_2 and $\dot{\Theta}_2$
versus time from $\Theta_2 = 0$ to $\Theta_2 = 180$ deg.

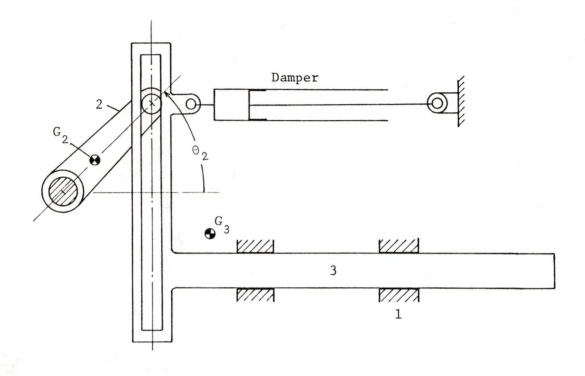